Women in Engineering and Science

Series Editor

Jill S. Tietjen, Greenwood Village, USA

The Springer Women in Engineering and Science series highlights women's accomplishments in these critical fields. The foundational volume in the series provides a broad overview of women's multi-faceted contributions to engineering over the last century. Each subsequent volume is dedicated to illuminating women's research and achievements in key, targeted areas of contemporary engineering and science endeavors.The goal for the series is to raise awareness of the pivotal work women are undertaking in areas of keen importance to our global community.

Anne Lucietto
Editor

Innovating for Tomorrow: Engineering Technology

Stories from Faculty, Students, and Industry Professionals

Springer

Editor
Anne Lucietto
Purdue University
West Lafayette, IN, USA

ISSN 2509-6427 ISSN 2509-6435 (electronic)
Women in Engineering and Science
ISBN 978-3-032-07644-1 ISBN 978-3-032-07645-8 (eBook)
https://doi.org/10.1007/978-3-032-07645-8

This Springer imprint is published by the registered company Springer Nature Switzerland AG
The registered company address is: Gewerbestrasse 11, 6330 Cham, Switzerland

Dedication

To all those who encouraged, mentored, guided, advised, and supported us—this book is for you.

To the teachers who lit the spark, the advisors who helped us find our way, the mentors who offered wisdom, and the friends and family who stood beside us with unwavering belief your impact is woven into every page of this work.

And to our fellow contributors many of whom are the educators, advisors, and mentors we honor thank you for your dedication to shaping futures, challenging minds, and opening doors. Your commitment to teaching and guiding others is the foundation and future of engineering technology education.

This book is a tribute to our collective journey and our shared responsibility to support those who follow. May it inspire continued growth, connection, and transformation—for ourselves, our students, and the generations to come.

Foreword 1

It is with great pride and admiration that I introduce *Innovating for Tomorrow: Engineering Technology*. This timely and inspiring work celebrates the achievements of women in a field that is both foundational to our modern world and ripe for transformation.

Engineering technology has long been a cornerstone of innovation, bridging the gap between theoretical engineering and practical application. Yet, for too long, the field has not fully reflected the diversity of talent and perspective that exists in our society. This book is a powerful step toward changing that narrative.

Despite progress, women remain significantly underrepresented in engineering and engineering technology. In 2023, women earned only **22% of degrees in these fields** in the United States. Within the broader STEM workforce, women held just **35% of jobs**, and their presence in leadership roles is even lower. These numbers highlight the urgent need to create more inclusive pathways into these fields—not only to promote equity but also to address the growing demand for skilled professionals in an increasingly technology-driven world.

The stories and insights shared across its three sections—senior faculty and administration, faculty at all levels, and students and graduates—offer a compelling portrait of the women shaping the future of engineering technology. From pioneering educators and researchers to emerging leaders and passionate students, these voices highlight the profound impact women are making in classrooms, laboratories, and industries worldwide.

Equally important is the book's historical context and its examination of the many facets of engineering technology as a discipline. These chapters remind us that progress isn't just about increasing numbers but also about fostering inclusive environments where everyone can thrive, contribute, and lead.

At Purdue University's Polytechnic Institute and across the broader academic and professional landscape, we recognize that the future of engineering technology hinges on our ability to attract, support, and empower a diverse group of innovators. Encouraging more women to pursue degrees and careers in this field is not just about equity—it is a strategic necessity for addressing the complex challenges of tomorrow.

This book is both a celebration and a call to action. It honors the trailblazers who have paved the way and encourages the next generation to follow in their footsteps—and to create new paths of their own. I plan to purchase copies of the book and share it with colleagues and students.

I praise my colleague and the contributors to this volume for their dedication, vision, and commitment to promoting women in engineering technology. May their stories inspire curiosity, courage, and transformation.

Emeritus Dean and Distinguished Professor
Purdue University Polytechnic Institute,
West Lafayette, IN, USA
Gary Bertoline

Foreword 2

Women have been shaping the field of engineering technology for well over a century—often in ways that have transformed industries, expanded educational opportunities, and opened doors for future innovators. From the earliest pioneers who defied societal expectations to become engineers and technologists, to today's leaders driving advancements in automation, sustainable manufacturing, and applied engineering systems, women have continually redefined what is possible. Their impact extends across every branch of engineering technology, from product design and robotics, to advanced materials and energy systems, contributing not only technical expertise but also the leadership and vision required to solve complex, real-world problems.

This book, *Women in Engineering Technology*: *Innovating for the Future*, is organized into three sections: Senior faculty and administration, faculty at all levels, and graduates and students. Each section begins with an introduction highlighting the accomplishments and experiences of the women featured, providing a narrative thread that connects their individual stories to the broader history of the field. The historical chapter sets the stage, tracing the evolution of engineering technology as an academic discipline and professional practice. In contrast, the topical chapter on areas of study underscores the breadth of expertise from mechanical, electrical, and civil technology to cutting-edge specializations, such as additive manufacturing, mechatronics, and cybersecurity.

Research shows that visibility, mentorship, and representation play critical roles in inspiring women to enter and remain in technical careers. This is especially true in engineering technology, a discipline that bridges theoretical engineering concepts with practical application and innovation. The women highlighted in these pages—whether serving as deans, department chairs, faculty, industry leaders, or students—demonstrate the value of diverse perspectives in tackling challenges, such as sustainable infrastructure, digital transformation, and workforce development. Their work is not only advancing technology but also redefining who participates in creating it.

As we navigate a rapidly evolving technological landscape driven by automation, data analytics, clean energy, and advanced manufacturing, the question is not just “Where are the women in engineering technology?” but “How do we ensure their voices and innovations shape the future?” The stories in this book offer compelling answers. We see educators redesigning curricula to better align with industry needs, administrators promoting equitable opportunities, graduates applying their skills to entrepreneurial ventures, and students conducting research that addresses global challenges.

This book is both a celebration and a call to action. It celebrates the trailblazers whose achievements paved the way, the practitioners whose work defines excellence today, and the emerging professionals who will lead tomorrow’s breakthroughs. It also calls on institutions, industries, and communities to recognize that inclusive innovation is essential for progress.

Whether you are a student considering a career in engineering technology, a professional seeking inspiration, or a leader looking to foster a more diverse and dynamic workforce, you will find in these pages role models, mentors, and innovators whose stories resonate far beyond the lab or the classroom. I encourage you to not only read their stories but also share them—because the future of engineering technology will be stronger, more creative, and more impactful when it reflects the full spectrum of talent and perspective.

Department of Engineering Technology
Kennesaw State University
Kennesaw, GA, USA

Aaron Lee Adams

Preface

When we first envisioned this book, we weren't just compiling stories; we were building a platform for voices that have too often been sidelined in the narrative of engineering and technology voices of women who have not only contributed to the field but who have shaped it, challenged it, and redefined its future.

As an editor and contributor, I had the privilege of witnessing firsthand the depth and diversity of these experiences. Each chapter, interview, and reflection in this book is the result of countless hours of conversation, research, and collaboration. What emerged is not a single story but a vibrant tapestry of ambition, resilience, and innovation.

This book is for the student wondering if she belongs in a robotics lab, the mid-career professional navigating leadership in a male-dominated space, the educator, the policymaker, the ally—anyone who believes that engineering technology should reflect the full spectrum of human potential.

It is also a book about systems. About the structures that have been excluded and the ones we must now build to include. It asks hard questions and offers hopeful answers. It doesn't shy away from the challenges, but it also doesn't let them define the narrative.

To the women who shared their stories: thank you for your honesty, your brilliance, and your courage. To the readers: may you find in these pages not only inspiration but also a call to action.

Because representation is not the end goal it's the beginning of a better, more equitable future in engineering technology.

West Lafayette, IN, USA — Anne Lucietto

Acknowledgments

This book is the result of many hands, minds, and moments of collaboration. We are deeply grateful to our colleagues, students, and institutional partners who contributed their time, insights, and experiences to this work. Your openness, curiosity, and commitment to advancing engineering technology education have shaped every chapter.

We also recognize the role of digital tools and emerging technologies in supporting our process. Artificial intelligence assisted in organizing and refining transcripts, sorting qualitative data, and correcting grammar and spelling inconsistencies. These tools helped us maintain clarity and coherence while preserving the authenticity of our contributors' voices.

Thank you to the institutions that supported our work, the reviewers who offered thoughtful feedback, and the administrative teams that helped coordinate logistics. Your support made it possible to bring this work to life.

We acknowledge the spirit of collaboration that defines this field. Whether through mentoring, advising, teaching, or learning, each contribution reflects a shared commitment to building inclusive, innovative, and impactful pathways in engineering technology.

Finally, I would like to acknowledge the support provided by the *Purdue Books Initiative*.

Contents

About the Authors

Carol Richardson holds a Master of Science in Electrical Engineering from Union College and a Bachelor of Science in Electrical Engineering from the University of Wyoming. At the Rochester Institute of Technology (RIT), she has served in numerous leadership roles, including Interim Dean and Vice Dean of the College of Applied Science and Technology (CAST), Department Chair and Program Chair within the Electrical, Computer, and Telecommunications Engineering Technology (ECTET) department, and as a Professor. Her academic and professional focus has been on curriculum development, assessment, and the advancement of Engineering Technology education. A recognized leader in her field, Professor Richardson is a Fellow of both ABET, Inc. and the American Society for Engineering Education (ASEE). Her accolades include the ASEE Frederick Berger Award, the ASEE James McGraw Award, and the IEEE Accreditation Board Award for Meritorious Achievement.

Elizabeth (Betsy) Dell is a Manufacturing and Mechanical Engineering Technology Department Professor at the Rochester Institute of Technology. With a career spanning over three decades, she has significantly contributed to gender equity in STEM. As the Director of AdvanceRIT and Senior Faculty Associate to the Provost at RIT, she led initiatives to support women faculty, fostering an inclusive academic environment. Betsy's tenure as the Director of Women in Technology and her role as Faculty Associate to the Provost for Women Faculty underscore her commitment to advancing women in engineering and technology. Her research and publications often focus on strategies to recruit, retain, and advance women in STEM, including the development of professional development workshops and mentoring programs. Dell has been instrumental in securing numerous grants aimed at promoting gender diversity, such as the NSF ADVANCE PLAN D and the NSF INCLUDES WiSEN project. Her efforts have been recognized with several awards, including the NIH Prize for Enhancing Faculty Gender Diversity, RIT's Edwina Award, and Isaac L. Jordan Faculty Pluralism Award.

Angie Hill Price is a pioneering engineering technology leader whose career exemplifies the transformative power of mentorship, resilience, and inclusive leadership. From her early struggles as an underprepared undergraduate to her current role as Associate Dean at Texas A&M University, Dr. Hill Price's story offers a compelling blueprint for women navigating the intersections of technical expertise, academic leadership, and personal fulfillment. Her narrative challenges traditional STEM success models, highlighting the value of service, relational leadership, and work–life integration. As part of a broader volume on women in engineering technology, this chapter contributes a vital perspective on how women are not only entering the field but also reshaping it from the inside out.

Alisa Heinie offers a perspective rooted in academic governance and curriculum innovation. Her contributions focus on the intersection of policy, pedagogy, and accreditation, emphasizing how senior faculty and administrators must navigate institutional structures while advocating for the unique needs of engineering technology programs. She discusses the critical role of shared governance, faculty voice, and strategic planning in sustaining program relevance and quality.

Lesley Berhan 's journey from Trinidad to Toledo exemplifies resilience, adaptability, and a passion for education. After excelling in civil engineering and earning a master's degree at MIT, she transitioned from structural engineering to academia, earning a PhD in Mechanical Engineering from the University of Michigan. At the University of Toledo, Lesley has become a professor, expanding her role to include student success and diversity initiatives. Her research focuses on equity, access, and the experiences of both engineering and engineering technology students. Lesley's development of a pre-major program has provided alternate pathways for students who may not meet traditional admission criteria, significantly improving retention and success rates. Her work underscores the importance of understanding diverse pathways and providing equitable access to education, inspiring aspiring engineers and educators.

Anne Lucietto is an accomplished engineer, educator, and advocate with nearly 30 years of industry experience. During her time in industry, she also taught as an adjunct at a community college, sparking her transition into academia. Anne's career is rooted in a family tradition of craftsmanship, language, and creativity, blending technical expertise with a deep commitment to mentorship and community engagement. Dr Lucietto has held leadership roles in corporate and nonprofit sectors, including serving as treasurer for the Society of Women Engineers (SWE) and chairing her local planning commission. She modeled resilience and curiosity as a mother, sharing her professional journey with her daughter. Anne has inspired thousands of students through K–12 STEM outreach, speaking at major events, and celebrating academic achievement. Her legacy is empowerment, innovation, and dedication to creating a more inspired future for all learners.

Brittany Newell , raised on a farm in Southern Indiana, pursued her passion for science and math through a Biomedical Engineering degree at Purdue University. She earned both her master's and PhD in Agricultural and Biological Engineering, focusing on Biotechnology and Machine Systems. After working in industry as a quality engineer and regulatory manager, she returned to Purdue as a Professor in the School of Engineering Technology. Her research centers on embedded sensing systems, material optimization, low-cost manufacturing, and additive techniques. She also explores electroactive polymers, pneumatic actuators, and the integration of AI and machine learning in control systems and education. Brittany teaches thermodynamics and engineering technology fundamentals. Beyond her professional achievements, she is a dedicated mother of four, wife, and active contributor to her family farm, balancing technical innovation with personal fulfillment.

Andrea Fay Barber-Dansby 's forty-plus-year engineering technology journey reads like Dr. Seuss' book, *Oh the Places You'll Go*. Her industry and higher education experiences have contributed to a strong skill set she can utilize in academia and community building sectors. Fay is on a collective mission with others to increase the number of women enrolled in engineering technology and graduating with those degrees. Her undergraduate studies and affiliations at Purdue University and graduate studies at Columbia University led to her employment and a successful career with a component division of General Motors. Fay seeks out new opportunities and experiences beyond her job description. She encourages our next generation, hoping they will push aside fears and doubts and move forward in enrolling and graduating with that door opener degree in engineering technology.

Beth McGinnis-Cavanaugh is a civil engineer, educator, and passionate advocate for community colleges and women in STEM. A graduate of the University of Massachusetts Amherst, with bachelor's and master's degrees in Civil (Structural) Engineering, Beth began her academic journey at Springfield Technical Community College (STCC), where she now serves as a dedicated Professor. Her career has spanned teaching, curriculum development, advocacy of community college engineering pathways, and national STEM education initiatives, including multimillion-dollar NSF-funded projects. Known for her empathetic teaching style and commitment to equity, Beth is an active member of the Society of Women Engineers (SWE) and the American Society for Engineering Education (ASEE). Her work inspires students and educators alike, especially those navigating nontraditional paths into engineering.

Jérika Lamas, a Peruvian born in Brazil, grew up in a warm but disciplined household shaped by her parents' resilience and sacrifices. Encouraged to pursue her strengths, she attended a rigorous technical school, earning a degree in Electronics. Her mother insisted she learn English, a decision that would later open global opportunities. Inspired by her father and sister, both engineers, Jérika pursued Materials Science and Engineering at university, where she excelled academically and discovered a passion for research. After an eye-opening internship in Slovenia, she pursued a PhD in Belgium, following both love and academic opportunity. The transition was difficult—marked by cultural shock, isolation, and discrimination—but she persevered. Over time, she found her voice, stood up against prejudice, and embraced her identity. Her PhD in Solid State Science deepened her expertise and resilience. Jérika's story is one of courage, determination, and the pursuit of knowledge across borders.

Upinder Kaur From a young age, Dr. Upinder Kaur dreamed of becoming an astronaut. While other children played, she immersed herself in books about space, NASA, and the lives of astronauts, finding purpose in their stories. Growing up in Delhi, India, where space careers were rare—especially for girls—her ambition stood out. She discovered that many astronauts began as mechanical engineers, and this realization became her guiding light. To Upinder, mechanical engineering was more than a field of study—it was a launchpad to the stars, blending creativity, logic, and real-world impact. Despite knowing she would often be among the few women in the field, she embraced the challenge. Her journey reflects a passion for space and a deep resilience and determination to break barriers. Upinder's story is one of vision, courage, and the unwavering belief that even the most distant dreams can be reached with the right foundation.

Devon Pessler is an innovative engineering technology professional whose journey began with a childhood love for building with Lego bricks. Initially interested in architecture, she shifted to engineering technology in high school through hands-on learning. At Purdue University, she explored both Mechanical Engineering and Mechanical Engineering Technology (MET), ultimately choosing MET. Her academic path included developing a wearable medical device, researching acoustic guitar soundboards, and consulting for Purdue's Technical Assistance Program (TAP). With support from mentors like Mark French, Paul McPherson, and her father, Devon overcame personal and academic challenges with resilience. Now a visiting professor at Purdue, she teaches hands-on engineering courses and actively contributes to the Society of Manufacturing Engineers (SME) through the student chapter. Devon is passionate about empowering women in engineering technology and advancing the field through research and industry collaboration. Her story highlights the importance of mentorship, perseverance, and experiential learning in shaping a successful and impactful career.

Diana Narvaez 's commitment to inclusivity and support drives Diana Narvaez's journey as a PhD student in Mechanical Engineering Technology at Purdue University. As a woman in a male-dominated field, she understands the barriers underrepresented students face and works to create a welcoming environment. As a teaching assistant, she fosters respectful, engaging classrooms where every student feels heard and valued. Her experiences of isolation during college inspire her empathetic teaching style, which includes adapting to diverse learning needs and encouraging open dialogue. Diana is primarily dedicated to mentoring marginalized students, offering guidance, and sharing her story to help others navigate challenges. She believes inclusive spaces empower students to take risks, grow, and contribute meaningfully to the field. Her work reflects a deep belief that belonging strengthens both individuals and the future of engineering technology.

Colleen McGuiness, Esq, graduated from Purdue University with a degree in Mechanical Engineering Technology in 2021. While at Purdue, Colleen interned with a contractor for the Department of Defense and worked as an assistant student researcher. Upon graduation, Colleen worked at Caterpillar Inc. as a technical marketing representative in their Oil and Gas Industry. Colleen began to pursue a law degree at New York Law School in 2022 to become a patent lawyer. During her time at New York Law School, Colleen was an affiliate of the Innovation Center, Patent Committee Chair for the Intellectual Property Law Society, and a member of the Patent Clinic, where she helped independent inventors seek protection rights for their inventions. Colleen is now a licensed Intellectual Property Attorney, assisting innovators to protect their ideas, and graduated from New York Law School in May 2025.

Julia Rowe, MSET, has a passion for math and science that began in high school, where she discovered a love for problem-solving and innovation. Her ability to break down complex problems into manageable parts naturally led her to pursue engineering technology. Julia was drawn not only to the theory but also to the real-world applications of STEM, finding joy in hands-on learning and practical solutions. During her undergraduate studies, Julia faced both rewarding and challenging experiences. The rigorous coursework tested her resilience, especially in adapting to traditional teaching methods. However, she embraced these challenges, learning to persevere and grow. Hands-on labs became a highlight, helping her grasp difficult concepts and reinforcing her belief in experiential learning. Julia's journey was one of self-discovery and determination. She realized her skills could be used to make a meaningful impact, inspiring her to apply engineering principles to benefit communities. Her story reflects resilience, adaptability, and a passion for purposeful innovation.

Meher Taleyarkhan, MSET 's fascination with roller coasters during her childhood sparked her interest in engineering, leading her to pursue mechanical engineering technology at Purdue University. Despite initial setbacks, Meher found success in the School of Engineering Technology, engaging in undergraduate research and improving her academic performance. She later pursued a master's degree, conducting a financial and engineering analysis of renewable energy systems. As a systems engineer at Northrop Grumman, Meher integrates diverse engineering disciplines to achieve cohesive project outcomes. Her ability to identify gaps and implement innovative solutions has made her an invaluable asset to her team. Meher is committed to mentoring junior engineers and advocating for gender diversity in the workplace. She actively participates in initiatives aimed at supporting and empowering women in engineering. Meher aims to advance into technical leadership positions, support and mentor young engineers, contribute to sustainable solutions, and balance technical and business acumen. Her vision for the future includes continuous learning and fostering a diverse and inclusive engineering community.

Olivia Engvall 's journey into engineering began with her involvement in her high school's FIRST Robotics team, sparking her interest in mechanical engineering. She pursued a degree in Mechanical Engineering Technology at Purdue University, complemented by a minor in Spanish, which led to a transformative year studying abroad in Lima, Peru. During that time, she received a degree in Mechanical Engineering from Universidad de Ingeniería y Tecnología. Olivia's career development was influenced by internships at Amway, where she now works as a project engineer. Her role involves supporting the launch of new products and day-to-day operations. Olivia values the collaborative environment at Amway and the mentorship she receives, which has been crucial in overcoming challenges and enhancing her problem-solving skills. Enthusiastic about encouraging young women to pursue engineering, Olivia emphasizes the importance of confidence, creativity, and continuous learning. She is involved in initiatives like FIRST Robotics and professional development groups.

Mica Caddick, MSML The journey for Mica Caddick, MSML, into STEM began in an unexpected place: a high school English classroom. Once set on teaching literature, she never imagined a future filled with CAD software, mechanical systems, and production workflows. But a chance encounter with hands-on technical work sparked a new passion rooted in problem-solving and real-world application. Transitioning into engineering technology wasn't easy. Mica faced steep learning curves and the challenge of entering a male-dominated field. Yet, with the support of mentors and her own determination, she found confidence and purpose. Her work in engineering technology revealed not only her technical strengths but also a new calling: education. Today, as a STEM educator in Indiana, Mica blends her love of language with her technical expertise, inspiring young learners to embrace curiosity and resilience.

Her story proves that STEM isn't a straight path—it's a dynamic journey where every background has value.

Selena Nadjarian, a Lebanese-Armenian mechanical engineering student born in California, has built a life at the intersection of cultures, disciplines, and resilience. Once set on becoming a doctor, she found her true passion in engineering—drawn to its logic, creativity, and potential for impact. Her journey includes surviving bullying, enduring war, and overcoming academic and cultural barriers. Through internships and collaborations with engineering technology professors, Selena discovered a love for education—designing interactive tools that make STEM accessible and intuitive for young learners. From gamified dam simulations to a culturally adapted walker inspired by her grandmother, her work reflects empathy and purpose. Therapy, mentorship, and mindfulness have helped her grow into a quiet leader, mentoring women in STEM and advocating for inclusive design. Selena's story is one of persistence and compassion—a reminder that engineering is not just about solving problems but building futures with heart and intention.

Part I
Foundations of Engineering Technology

Engineering technology (ET) emerged in the mid-twentieth century as a distinct educational pathway designed to bridge the gap between theoretical engineering and hands-on technical applications. Rooted in practical problem-solving and implementation, ET programs were developed to meet the growing demand for professionals who could apply engineering principles directly in industrial and technological settings. These programs emphasize applied learning, laboratory experience, and industry-relevant skills, preparing graduates for design, development, testing, and manufacturing careers.

Over time, the need for consistency and the development of academic standards led to the involvement of accreditation bodies. ABET, Inc. (formerly the Accreditation Board for Engineering and Technology), through its Engineering Technology Accreditation Commission (ETAC), ensures that ET programs meet similar standards through curriculum, faculty qualifications, facilities, and continuous improvement criteria. ABET's accreditation standards and oversight validate the quality of an ET program and further enhance graduates' employability and eligibility for professional licensure in many jurisdictions.

This part explores the evolution of Engineering Technology education and the critical role ABET has played in shaping and maintaining its standards.

A Brief History of Engineering Technology

Anne Lucietto

Engineering technology (ET) is an important academic field that connects theoretical engineering with practical technical applications. Unlike traditional engineering, which focuses on conceptual design, advanced mathematics, and scientific theory, engineering technology emphasizes hands-on application of engineering principles in real-world situations. This major appeals to students who enjoy practical problem-solving, using tools and systems, and gaining job-ready skills upon graduation.

The emergence of engineering technology as a distinct major reflects the changing needs of industrial society. As industries became more complex during the Industrial Revolution and into the twentieth century, there was a growing demand for professionals who could implement and maintain the innovations developed by engineers. This required a different educational background focused not just on science and mathematics but also on practical application and systems integration.

In the mid-twentieth century, engineering technology began to gain academic recognition as a separate field, particularly in the United States. Educational institutions started to offer specialized programs that combined theoretical study with hands-on laboratory experience, preparing students for careers in manufacturing, electronics, construction, and new technologies. These programs evolved into comprehensive degree offerings at the associate and bachelor's levels, eventually expanding to include master's and doctoral degrees.

Today, engineering technology is available at hundreds of colleges and universities across the nation and in various formats globally. In the U.S., several institutions have significantly influenced the development of the field's curriculum and professional standards. For example, Purdue University offers a range of

A. Lucietto (✉)
Purdue University, West Lafayette, IN, USA
e-mail: lucietto@purdue.edu

A. Lucietto (ed.), *Innovating for Tomorrow: Engineering Technology*, Women in Engineering and Science, https://doi.org/10.1007/978-3-032-07645-8_1

engineering technology degrees through its School of Engineering Technology within the Polytechnic Institute and is known for its leadership in applied engineering education. The Rochester Institute of Technology (RIT) provides advanced mechanical, electrical, and computer engineering technology programs. Texas A&M University boasts one of the largest and most comprehensive engineering technology departments. The University of Alabama in Huntsville emphasizes innovation and collaboration with industry, while the University of Toledo has established strong connections with regional industries, focusing on hands-on, career-oriented education. Pennsylvania State University offers programs across multiple campuses, supported by solid industry partnerships.

Various higher education institutions offer engineering technology programs at the associate, bachelor's, master's, and doctoral levels. These programs cater to a diverse student body and equip graduates for various technical and leadership roles in the industry. Associate-level programs, commonly found in community colleges, emphasize foundational technical skills and workforce readiness. In contrast, bachelor's degrees build on this foundation with advanced courses in systems integration, applied research, and project management. Although less common, master's and PhD-level engineering technology programs are growing to meet industry demands for innovation, applied research, and technical leadership.

A critical component of the educational ecosystem is the network of matriculation and transfer agreements that facilitates student mobility between institutions. Many students begin their academic journey at community colleges and later transfer to 4-year universities to complete their bachelor's degrees in engineering technology. These pathways are supported by articulation agreements, dual-enrollment programs, and statewide transfer initiatives that aim to streamline credit recognition and reduce barriers to degree completion. Such programs are especially vital for nontraditional students, working professionals, and those seeking affordable entry points into technical education.

In addition to traditional universities and community colleges, engineering technology programs are increasingly offered through polytechnic institutions, technical institutes, and online platforms, further expanding access and flexibility. Graduates from these and other institutions are employed in industries ranging from aerospace and automotive to various aspects of energy and information technology, often serving as the crucial link between design and execution.

In the following sections, we will briefly explore the history of engineering technology, examining how this major has evolved in response to technological change, economic demand, and educational innovation. This evolution has positioned engineering technology as a cornerstone of modern technical education—one that is deeply interconnected with workforce development, institutional collaboration, and access to STEM careers.

The origins of engineering technology can be traced back to the Industrial Revolution, spanning the late eighteenth to early nineteenth centuries. This transformative era saw the rise of steam engines, textile machinery, and electrical systems, revolutionizing industrial production. As these technologies became central to economic growth, the demand for skilled technicians and mechanics surged. While not formally trained as engineers, these individuals possessed the practical expertise to operate, maintain, and improve complex machinery. Their contributions laid the foundation for a new class of technical professionals who applied engineering principles in tangible, hands-on ways.

In response to this growing need, technical institutes and trade schools began to emerge, offering structured training in mechanical, electrical, and civil technologies. These institutions provided practical education that complemented the theoretical focus of traditional engineering programs and served as precursors to modern engineering technology curricula.

Formalization in the Twentieth Century

As industrial complexity intensified in the early twentieth century, with the advent of automobiles, widespread electrification, telecommunications, and eventually computing, the necessity for a workforce trained in applied engineering principles became increasingly apparent. This led to institutionalizing engineering technology as a distinct educational and professional pathway, separate from conventional engineering.

During the early decades of the century, many technical high schools and junior colleges began offering post-secondary programs in industrial arts and applied mechanics. Institutions such as Dunwoody College of Technology, founded in 1914 in Minneapolis, and the Milwaukee School of Engineering (MSOE), established in 1903, were pioneers in preparing students for immediate entry into technical fields. Their emphasis on hands-on learning and strong industry partnerships helped define the core values of engineering technology education.

Following World War II, the GI Bill dramatically expanded access to higher education, enabling millions of veterans to enroll in colleges and universities. This influx of students fueled a boom in technical education, transforming vocational schools into community colleges and elevating technical institutes into polytechnic universities. Institutions like Southern Polytechnic State University (now part of Kennesaw State University) and California Polytechnic State University (Cal Poly) broadened their offerings to include applied engineering disciplines, further legitimizing engineering technology as a vital academic field.

By the 1950s through the 1970s, the term "engineering technology" gained widespread use to describe programs focused on applying engineering principles rather than theoretical design. Purdue University played a leading role in this evolution by establishing its School of Technology—now the Purdue Polytechnic Institute, which offered degrees in mechanical and electrical engineering technology. Rochester Institute of Technology (RIT) and Pennsylvania College of Technology also emerged as leaders in providing 2-year and 4-year degrees tailored to industry needs.

A significant milestone occurred in 1979 when ABET (Accreditation Board for Engineering and Technology) created a separate commission to accredit engineering technology programs. This move formalized the distinction between engineering, accredited by the Engineering Accreditation Commission (EAC), and engineering technology, accredited by the Engineering Technology Accreditation Commission (ETAC). ABET accreditation ensured that engineering technology graduates met rigorous industry standards and could pursue professional certifications and licensure in certain states.

Defining the Characteristics of Engineering Technology Programs

Engineering technology programs are distinguished by their emphasis on laboratory work, applied mathematics, and real-world problem-solving. The curriculum is designed to prepare students for immediate entry into technical careers, focusing on practical skills and systems integration. Graduates often pursue roles as manufacturing or process engineers, field engineers, systems integrators, or technical project managers. Degree offerings typically include associate degrees (AAS), bachelor's degrees (BSET), and, increasingly, master's degrees in engineering technology, with doctorates in technology. As engineering technology evolves, its historical roots and educational innovations remain central to its identity. The discipline is a testament to the enduring value of applied knowledge in driving technological progress and meeting the demands of a dynamic industry-driven economy.

The Rise of Technical Education in the Twentieth Century

A turning point in the development of technical education occurred in the early 1900s, particularly in the United States and other industrialized nations. Building on the foundation laid by the Morrill Acts of 1862 and 1890, which established land-grant colleges with strong programs in engineering and agriculture, many institutions began to expand their offerings to include applied sciences and technologies. Over time, these colleges introduced 2-year associate and 4-year bachelor's degrees, emphasizing practical, career-oriented education. Graduates from these programs gradually moved into positions requiring technical expertise and leadership skills.

By the mid-twentieth century, the term engineering technology gained traction, especially in the United States. The post-World War II economic boom led to rapid growth in manufacturing, aerospace, and electronics, creating an increased need for professionals who could apply engineering principles in practical contexts. This demand resulted in the formalization of engineering technology as a separate academic track, distinct from traditional engineering programs, with a focus on practicality and technical skills.

A significant milestone in the evolution of engineering technology came with the establishment of accreditation standards by ABET (Accreditation Board for Engineering and Technology) in the 1960s. Acknowledging the unique goals of engineering technology education, ABET developed criteria for accrediting these programs. This helped solidify the identity of engineering technology as a legitimate academic discipline, emphasizing hands-on instruction, applied mathematics and science, industry-standard tools, and project-based learning as core components.

Graduates of engineering technology programs commonly pursued careers in product design, manufacturing, quality assurance, and systems integration. Many also sought professional certifications through organizations like the National Institute for Certification in Engineering Technologies (NICET), enhancing their credentials and career opportunities.

As globalization and technological advancements accelerated in the late twentieth and early twenty-first centuries, engineering technology programs expanded beyond the United States. Countries such as Canada, Germany, India, and China developed tailored models of applied engineering education to meet the needs of local industries. This global expansion highlighted the recognition of engineering technology as vital to technological advancement and economic growth.

This ongoing evolution reflects the adaptability and significance of engineering technology education. Rooted in the practical demands of the Industrial Revolution and shaped by decades of innovation, the field continues to equip students for dynamic careers at the intersection of engineering, technology, and industry.

The Future of Engineering Technology

Engineering technology is a crucial part of the broader STEM system today. As industries increasingly rely on automation, data analytics, and interdisciplinary collaboration, the role of engineering technologists has become more important than ever. These professionals are positioned to bridge the gap between conceptual innovation and practical implementation, ensuring that emerging technologies are effectively integrated into real-world systems.

Several key trends are shaping the future of engineering technology, with the integration of artificial intelligence and machine learning into engineering processes being one of the most transformative. Engineers trained in these areas work alongside intelligent systems to optimize manufacturing, enhance predictive maintenance, and streamline operations across various sectors. Additionally, there is a growing focus on green technologies and sustainability. Programs are evolving to include coursework and research in renewable energy, sustainable manufacturing, and environmental systems, reflecting the global push toward more responsible and resilient infrastructure.

The educational landscape is also experiencing significant change. Online and hybrid learning models are expanding access to engineering technology programs, allowing students to gain practical skills through virtual labs, simulations, and remote collaboration tools. This shift makes technical education more flexible and inclusive, accommodating diverse learning styles and life circumstances.

Another defining feature of the future is the deepening collaboration between academia and industry. Institutions are building stronger partnerships with companies to ensure that curricula remain aligned with workforce needs. These collaborations often lead to co-developed courses, internship opportunities, and joint research initiatives that prepare students for immediate career impact.

The history of engineering technology reflects a dynamic interplay between education, industry, and innovation. From its origins in the Industrial Revolution to its current role in shaping the future of technology, engineering technology has continually adapted to meet the demands of a changing world. As it evolves, it empowers students to become hands-on problem-solvers—professionals who understand technology and know how to apply it in meaningful, transformative ways (Brunner, 1964; Buchananm, 2019; O'Hair, 1995).

References

K. A. Brunner. History of engineering technology. Journal of Engineering Education **54**, 398–399 (1964)

W, W., m. Buchananm. A Brief History of Engineering Technology and a Case for Applied Engineering 2018 Gulf Southwest Section Conference, AT&T Executive Education and Conference Center, Austin, TX 78705 (2019)

M. O'Hair, (Ed.). Engineering Technology: an ASEE History. Oregon Institute of Technology (1995)

Accreditation and the Evolution of Engineering Technology Education

Anne Lucietto

Accreditation is a cornerstone of quality assurance in higher education, particularly in technical and professional disciplines where consistency, rigor, and industry relevance are essential. In the United States, ABET, the Accreditation Board for Engineering and Technology, has long been recognized as the gold standard for accrediting engineering and engineering technology programs. This chapter explores ABET's origins and evolution, its influence on engineering technology education, and the distinctions it draws between engineering and engineering technology curricula.

ABET traces its origins to 1932, when it was founded as the Engineers' Council for Professional Development (ECPD). It was established to standardize and unify engineering education in response to the rapid industrialization of the early twentieth century. As engineering disciplines diversified and the demand for technically skilled professionals grew, a structured, peer-reviewed accreditation process became imperative. ECPD provided that structure, ensuring engineering programs across institutions met consistent academic and professional standards.

In 1980, ECPD was renamed ABET, and its scope expanded to include engineering and engineering technology, computing, and applied science programs. This shift acknowledged the growing importance of hands-on, application-focused education alongside traditional theoretical engineering.

The emergence of engineering technology as a distinct academic discipline in the mid-twentieth century was driven by industry's need for professionals who could bridge the gap between theoretical engineers and hands-on practitioners. Engineering technology programs emphasize practical applications over abstract theory,

A. Lucietto (✉)
Purdue University, West Lafayette, IN, USA
e-mail: lucietto@purdue.edu

A. Lucietto (ed.), *Innovating for Tomorrow: Engineering Technology*, Women in Engineering and Science, https://doi.org/10.1007/978-3-032-07645-8_2

preparing graduates to implement, operate, and improve existing technologies rather than design new ones from first principles.

These programs emphasize laboratory-based instruction, applied mathematics and science, industry-standard tools, and collaborative, project-based learning. Graduates often pursue product design, manufacturing, quality assurance, and systems integration careers. Many also seek professional certification through organizations such as the National Institute for Certification in Engineering Technologies (NICET), further validating their expertise and enhancing their professional standing.

Recognizing this distinction, ABET established the Engineering Technology Accreditation Commission (ETAC), a separate body responsible for accrediting engineering technology programs. This allowed for the development of tailored accreditation criteria that reflect the unique goals, learning outcomes, and career pathways associated with engineering technology education.

ABET has played a critical role in legitimizing engineering technology as a rigorous and respected academic discipline. Through its evolving accreditation framework, ABET ensures that programs meet high standards of quality and relevance, shaping the future of technical education and supporting the development of a skilled, adaptable workforce.

ABET Accreditation: Purpose and Process

As engineering technology matured into a distinct academic discipline, the need for formal quality assurance became increasingly important. ABET established a voluntary, peer-reviewed accreditation process to evaluate educational programs against rigorous standards. This process assesses several key areas, including curriculum alignment with industry needs, faculty qualifications and engagement in continuous improvement, the adequacy of facilities such as laboratories and classrooms, student learning outcomes, and the program's commitment to ongoing development through data-driven decision-making.

ABET's Engineering Technology Accreditation Commission (ETAC) emphasizes the practical application of engineering principles for engineering technology programs. Programs are expected to demonstrate that graduates possess strong technical skills, can work effectively in teams, communicate clearly, and solve problems in real-world contexts. The goal is to ensure that students are technically competent and adaptable to the industry's evolving demands.

Engineering vs. Engineering Technology: Curriculum Differences

Although engineering and engineering technology programs fall under the ABET umbrella, they differ significantly in educational philosophy and curricular structure. Traditional engineering programs, accredited by the Engineering Accreditation Commission (EAC), focus on theoretical analysis and design. These programs typically require extensive calculus, differential equations, and advanced physics coursework and often emphasize simulation-based labs and research-oriented learning.

In contrast, engineering technology programs, accredited by ETAC, prioritize application and implementation. The curriculum leans more heavily on algebra, trigonometry, applied calculus, and practical sciences. Students engage in extensive hands-on laboratory work, using industry-standard tools and techniques to solve tangible problems. Career pathways also diverge: engineering graduates often pursue roles in design engineering, research and development, or graduate studies, while engineering technology graduates typically enter the workforce as field engineers, engineers, operations specialists, or manufacturing professionals.

These distinctions reflect the complementary nature of the two fields. Engineers are often responsible for conceptualizing and designing systems, while engineering technology graduates, also referred to as engineers, ensure those systems are built, maintained, and optimized for performance in real-world environments. Together, they form a dynamic partnership that drives innovation and operational excellence across industries.

The Impact of ABET on Engineering Technology

The establishment of ABET accreditation has played a transformative role in legitimizing engineering technology as a rigorous and respected academic discipline. ABET has helped standardize program quality across institutions through its structured evaluation process, ensuring consistency in educational outcomes and alignment with industry expectations. This standardization has significantly improved the recognition of engineering technology degrees among employers and professional organizations, enhancing the credibility of graduates in the workforce.

In some states, ABET accreditation has also enabled engineering technology graduates to pursue professional licensure, further expanding their career opportunities. Additionally, by offering an alternative pathway into STEM fields, engineering technology programs have contributed to greater diversity in engineering education. These programs often attract students from varied backgrounds who are drawn to the hands-on, application-oriented nature of the curriculum.

ABET's continuous improvement model has been instrumental in keeping engineering technology programs responsive to technological change. Institutions are encouraged to regularly assess and update their curricula to reflect emerging trends such as automation, sustainability, and digital manufacturing. This dynamic approach ensures graduates have relevant skills and knowledge to meet the industry's evolving demands.

Ultimately, ABET's role in accrediting engineering technology programs has shaped the field's identity and credibility. By recognizing the distinct contributions of engineering technology graduates, ABET has helped build a more inclusive and responsive engineering education ecosystem. As technology advances, the standards and expectations for those who design, build, and maintain the systems that power our world will likewise evolve—guided by the foundational principles of quality, relevance, and continuous improvement (ABET, 2015; 2016; 2026).

References

ABET. *Engineering vs. Engineering Technology* (2015) http://www.abet.org/accreditation/new-to-accreditation/engineering-vs-engineering-technology/

ABET. *Setting the Standard World Wide* (2016) http://www.abet.org/accreditation/

ABET. *ABET Home Page*. (2026) http://www.abet.org

Part II
Senior Faculty/Administration in Engineering Technology

The landscape of engineering technology education is shaped not only by curriculum and industry alignment but also by senior faculty and academic administrators' vision, leadership, and strategic direction. As institutions face increasing demands for innovation, equity, and workforce readiness, the role of experienced educational leaders becomes central to navigating these complexities and ensuring programmatic excellence.

This part, *Senior Faculty and Administration in Engineering Technology*, brings together the voices and experiences of four distinguished leaders in the field: Prof. Carol Richardson, Prof. Elizabeth (Betsy) Dell, Dr. Angie Hill Price, and Dr. Alisa Henrie. Each author offers a unique perspective on the evolving responsibilities of senior faculty and administrators, grounded in decades of service, scholarship, and leadership within engineering technology programs.

These chapters provide a comprehensive view of what it means to lead in engineering technology today. They underscore the multifaceted nature of academic leadership, where vision, collaboration, and resilience are essential. Whether managing accreditation cycles, mentoring junior faculty, or responding to shifting industry demands, senior faculty and administrators are the stewards of tradition and transformation.

This part invites readers to engage with these leaders' lived experiences, reflect on their roles within academic institutions, and consider how leadership in engineering technology can continue to evolve in service of students, faculty, and society at large.

A Career in Engineering Technology at Rochester Institute of Technology

Carol Richardson

Introduction

The National Academy of Engineering (NAE) did a study of Engineering Technology (ET) education in the United States in 2017 and found that a discussion of the role that it should play in supporting the nation's technical infrastructure is often absent from pertinent discussions.[1] There are a substantial number of credentialed ET professionals working across the field of engineering. This chapter describes the academic career of a woman faculty member, Carol Richardson, who spent over 30 years developing ET programs at RIT in Rochester, New York. Most notably during this career, Professor Richardson and her colleagues designed and enacted what would become the two main recommendations of the 2017 NAE study, four decades before the study existed.

Background and Context

I began my career in Engineering Technology (ET) at RIT in 1978 as an adjunct professor in the Department of Electrical Engineering Technology (EET). I moved to the Rochester, New York area the prior year with a husband and a 2-year old boy, and I had a baby girl 2 months after I arrived. Prior to our move to Rochester, which

[1] National Academy of Sciences, Engineering Education in the United States, 2017.

C. Richardson (✉)
Rochester Inssitute of Technology, Rochester, NY, USA
e-mail: carite@rit.edu

A. Lucietto (ed.), *Innovating for Tomorrow: Engineering Technology*, Women in Engineering and Science, https://doi.org/10.1007/978-3-032-07645-8_3

we undertook for my husband's new job, I worked at General Electric as an instrumentation and control engineer for the Navy nuclear submarine program. My first position in industry after graduating with my Electrical Engineering (EE) degree from the University of Wyoming was at Collins Radio Company in Cedar Rapids, Iowa, where I was a design engineer for radio frequency radio equipment in the Telecommunications Systems Division. RIT advertised an adjunct faculty position in the local newspaper. I earned my master's degree in EE from Union College immediately before my son was born, but I had no teaching experience. I worked full-time for General Electric during the day and took my graduate courses each afternoon at 4:00. In 1978, RIT required ET faculty members to hold a master's degree in engineering and have at least 3 years of industrial experience to teach courses. These requirements remained in place for many years. I was given a textbook and a course outline and expected to design my own course without any help or mentoring. The course was titled Communication Systems II, and since I had industrial experience from Collins Radio Company with high-frequency radios, I was able to design a course.

Several faculty members in that department had professional leave the following year, so the department offered me a 1-year contract to teach full-time as a visiting professor. There were not any openings the subsequent year when everybody returned from leave, so I moved to RIT's College of Engineering (COE), since they had faculty members about to take their leaves. The Electrical Engineering Technology and Electrical Engineering departments were located on one floor of the Gleason Building, which was the home of the whole college. EET and Mechanical Engineering Technology shared laboratories and equipment with the larger college. I was the only woman faculty member in either department on the third floor at that time in the late seventies. The college leadership told me that if I wanted to continue teaching in EE, I would have to start working on my PhD. A local university, University of Rochester, had such a doctoral program, but they were known for making it difficult for RIT professors to advance. Instead, most RIT COE professors who were working on PhDs drove 60 miles to Buffalo, New York, to obtain their degree from the State University of New York at Buffalo. Childcare was not widely available in the late seventies, and I did not think that I could manage commuting to Buffalo with two children under 4 years of age. My childcare was provided by women coming to my home. The Department of EET in the College of Applied Science (CAST) at RIT had an opening for a visiting professor after I finished my second year of teaching at RIT. I accepted the offer and began my career teaching ET.

Personal Experiences and Stories

I had never heard of ET programs when I taught my first EET course in 1978. I spent the next 17 years teaching and learning about it. I was impressed by the ET curriculum, which was different from my engineering education, and by the support for students seeking positions upon graduation, support which I had not experienced.

At that time, ET faculty members at RIT usually did not have research projects; instead, we often consulted for industry. Space for our programs was limited, and there were not many research labs in either college. Professional leaves of ET faculty members were spent working with industry, so we could keep our courses current and develop new programs. External funding, which was useful for promotions and salary increases, was often obtained with equipment grants from local industry or the National Science Foundation (NSF). Student scholarships for ET students were also available from NSF.

RIT's first ET programs were all upper-division programs that were housed in the School of Engineering Technology (SET). Most of our students came from the SUNY community colleges. These students received associate degrees in ET and would go on to technician positions after graduating. Faculty members in the SET developed relationships with faculty at community colleges by way of an organization called the Engineering Technology Association, which was local to the state of New York. We met a couple of times a year at community colleges or RIT. RIT's SET faculty members made trips to the community colleges to recruit students for the RIT upper division ET programs. The ET and Engineering programs at RIT required a full year of cooperative education work experience before a student could graduate from the program. RIT provided much support to ET students to find these co-op positions. Parents began telling us that they did not want to send their children to a community college for 2 years and then to RIT, so we expanded all ET programs to include the first 2 years in-house at RIT. Adult students at the community colleges expressed a preference for the availability of a complete 4-year Engineering Technology program within the community colleges. They highlighted that their full-time employment and family commitments in the local area made relocating to pursue a bachelor's degree impractical. Several community colleges in New York implemented 4-year ET programs in response to their student needs.

Many of RIT's ET faculty members became active in local professional engineering organizations to develop consulting positions, obtain equipment donations, and promote ET. This activity was encouraged by the SET for the equipment donations and the creation of co-op positions in the local community for ET students.

I participated in both local and national engineering organizations. I was the first woman chair (1989) of the Rochester section of the Institute of Electrical and Electronic Engineers (IEEE), which had 1500 members at that time. I am now a life member of IEEE. I was the first woman president of the Rochester Engineering Society (RES), and this organization had 600 members at that time.

The equipment donations I was able to secure through my professional networks and the papers I was able to publish on these activities helped me achieve academic promotions and provided well-equipped laboratories for the department in the telecommunications program. RIT received $217,000 of equipment donations for telecommunications laboratory equipment before I became an administrator and $330,000 after I became department chair. These equipment donations came from a local Rochester telecommunications company and the NSF. In my leadership roles with IEEE and RES, I wrote the criteria for both ET and Engineering student scholarships. I presented papers on these activities at ASEE national conferences and Accreditation Board for Engineering and Technology (ABET) symposiums.

There was a great effort toward promoting Engineering Technology as a field at the national level. Much of this work was accomplished by the American Society for Engineering Education (ASEE), the ASEE Engineering Technology Division (ETD), and the ASEE Engineering Technology Council (ETC). One of the key tasks was strategizing how to educate those engineering companies that employed ET graduates about the field, professional engineering licensure, and federal job classifications. ETD and ETC worked on these issues for many years. The other national organization working on these issues was ABET, Inc. ABET was originally known as the accreditation board for Engineering and Engineering Technology undergraduate academic programs. This organization changed its name to ABET, Inc. when they began to accredit other undergraduate programs such as Computer Science and Information Technology. During this period, significant progress was made on the work of informing companies about ET graduates, so ET programs continued to expand. Progress on professional engineering licensure was made after I retired from RIT.

I was the first woman chair of the ETD of ASEE and a member of the ETC of ASEE. I also served as ASEE's Vice President of Professional Interest Councils and as a member of the ASEE executive board from 2007 to 2009. I was an IEEE program evaluator for ABET from 1994 to 1998 and an ABET commissioner for IEEE in 2001. I was the first woman chair of the Engineering Technology Accreditation Commission (ETAC) of ABET, and I oversaw the organization's name change from "Technology Accreditation Commission" to ETAC. These activities earned me several awards: an ABET Fellow, an ASEE Fellow, the ASEE Frederick J. Berger Award, the ASEE James H. McGraw Hill Award, and the IEEE Educational Activities Board Meritorious Achievement Accreditation Activities award.

My connections with national telecommunications organizations such as the International Communications Association (ICA) in the 1980s allowed me to design and propose a Bachelor of Science program that was the first Technology Accreditation Commission-accredited program in ABET. I received the ICA Leon Montgomery Academic Fellow Award and an equipment grant from the National Communications Association in 1989.

I accepted my first administrative assignment in the School of Engineering Technology (SET) in 1989 as the program chair of the Telecommunications Engineering Technology (TET) program, and then I wrote the proposal to New York State for the TET graduate program in 1993.

I became chair of the Department of Electrical, Computer, and Telecommunications Technology (ECTET) in 1994. Due to cost-cutting measures, the School of Engineering Technology was eliminated, and I reported directly to the dean of the College of Applied Science and Technology (CAST). The ECTET department at that time had approximately 400 students, 19 full-time faculty members, 16 adjunct faculty members, and 5 staff members. ECTET supported one associate degree program, three Bachelor of Science degree programs, and one graduate program, all in ET. I continued to work hard to secure equipment donations as the department chair and obtained $330,000 of equipment for the telecommunications program laboratory and $396,000 in scholarship funds for ET and Engineering students from the NSF.

CAST had never had a dedicated building, and we set out to fix that. CAST's large programs were ET, Computer Science, Hospitality, and Packaging Science. Each of these programs and the Dean's office were in a different building at RIT. Engineering Technology and Engineering programs were growing along with Computer Science and Information Technology. Space was always an issue between COE and CAST. The university decided to move some of the ET programs into a new space, and a building was constructed very quickly. Several of the programs moved into the new space in the 1990s. The Department of Information Technology in CAST was established and rapidly expanded during this period. CAST needed more space, so two different university development projects began. One project developed the Golisano College of Computing to manage Computer Science, Information Technology, and associated programs. The second project provided funding for an ET building for some of the ET programs. We centered the telecommunications ET program in our winning application for a $2,000,000 grant from the McGowan Foundation, named after William McGowan, the founder of the large telecommunications company, MCI. I spent a lot of time working on these funding projects as department chair.

In the 1990s, a representative from Project Lead the Way (PLTW) contacted RIT about developing a high school engineering program. This precipitated an ongoing connection between ECTET and high school technology teachers and students. The goal of PLTW is to increase the number of scientists, engineers, and technicians produced by our educational system. One of my faculty members worked with a technology teacher from a local school system in his digital logic freshman course. PLTW has grown from an engineering-only high school program to a program offering pathways in computer science, engineering, and biomedical science with 14,000 programs in over 11,500 schools. Many high school teachers have come to RIT and similar institutions for training on creating and running such programs, and, as a result, high school teachers, administrators, counselors, and students at large have become familiar with ET programs in the past 25 years.

I have always supported the women's movement that started when I was graduating from college. I was a charter member and past president of the Rochester Section of the Society of Women Engineers (SWE), and I worked with the RIT student chapter of SWE that we started at RIT many years ago. As a department chair, I received a grant of $26,000 in 2003 that allowed the college to begin a Women In Technology (WIT) program. This program is still active at RIT and is doing well. RIT Institution research shows that the percentage of women students in 2024–25 in the CET is 23.6% and that this figure marks a significant increase since RIT received the grant.

My next administrative assignment as Vice Dean of CAST began in 2005. I completed a great deal of fundraising work for the new RIT CAST Engineering Technology Building, which was awarded Gold Certification by the U.S. Green Building Council's Leadership in Energy and Environmental Design (LEED). This was the first LEED building at RIT due to the leadership of the CAST Department of Civil Engineering Technology faculty members in the design process of the building.

After serving as the Vice Dean of CAST, I became Interim Dean from 2007 to 2009. The new CAST ET building was finished in 2009, and I retired from RIT in 2010. RIT Institutional Data show that ET enrollments have steadily increased since then, and the ET program with the largest enrollment of students is Mechanical Engineering Technology.

Analysis and Discussion

In 2017, the National Academy of Engineering (NAE) published the results of an NSF-sponsored, 2-year study titled "Engineering Technology Education in the United States.".[2] NAE's charge was to shed light on the status, role, and needs of ET education in the United States. The study stated that "The number of degrees in Engineering Technology, while smaller than in engineering, is substantial." The study had five findings and associated recommendations. NAE recommended that two of the findings needed more research and one finding needed better data collection. NAE's other two findings are as follows:

Recommendation 1: Within academia, it is critical for leaders of 2-year and 4-year ET programs to engage more meaningfully in discussions with leaders in post-secondary education about the similarity and differences between the two variants of engineering and how they may complement one another while serving the interests of a diverse student population. This engagement can be accomplished in dialog within and between individual institutions; through work by discipline-based and affinity engineering professional societies; and by leaders within the American Society for Engineering Education, such as the Engineering Technology Council, the Engineering Dean's Council, and the Corporate Member Council.

RIT's ET programs have thrived, with the university rewarding ET faculty for their engagements with community colleges, professional engineering societies, and high schools through Project Lead the Way. It is interesting to observe that RIT began their engagement with community colleges in the 1970s and with high schools in the 1990s.

Recommendation 2: The ET education community should consider ways to make the field's value proposition more evident to K-12 teachers, students, students' parents, as well as to employers. Such an effort might include a new messaging developed in collaboration with a qualified public relations firm and based on data from market research on student and employer knowledge and perceptions of ET. The research might test the appeal and believability of rebranding ET as

[2] National Academy of Sciences, Engineering Education in the United States, 2017.

"applied engineering" or other appropriate names identified by the market research. Attention also should be paid to ways to reduce confusion associated with the term "Engineering Technology" and to simplifying the degree nomenclature. To encourage collaboration and avoid duplication, plans for any major new outreach should be communicated with appropriate leadership within the engineering education community, such as the Engineering Deans Council and Engineering Technology Council (ETC) of the American Society for Engineering Education.

RIT continues to work with students, parents, and employees of our graduates, and within our publications, to make the value proposition of ET more evident. RIT even changed the name of the College of Applied Science and Technology to the College of Engineering Technology recently.

The NAE study also discussed engineering licensure. The study stated that "In the United States, engineers must be licensed to perform certain tasks such as certifying the safety-related specification of design drawings." Individuals who have earned a 4-year engineering degree from an ABET-accredited program and who wish to be licensed must first pass the Fundamentals in Engineering (FE) exam, a test of broad knowledge in mathematics, science, and engineering. After gaining work experience (the amount of experience varies by state), those with the FE designation and requisite experience can take the "Principles and Practice in Engineering (PE) exam". Licensing is done by states, and in 2017, there were 30 states that allow those with an ET degree from an ABET-accredited program to take the FE and PE exams. In 2016, a professor from RIT, Martin Gordon, led an effort from the ASEE ETC to obtain equitable PE licensures for ABET/ETAC graduates. Today, there are only two states, Kansas and Illinois, where ABET/ETAC graduates cannot obtain a PE license although some of these states do require a MS degree. Federal job classifications need to be addressed, and this is an opportunity for the future direction of ASEE ET divisions.

Networking and faculty qualifications have evolved since I began my career in ET. The path that I used would need to be revised. Faculty members starting an academic career in ET now begin with larger networks and computers, and they can choose from a wider range of academic programs for their academic credentials. I had a satisfying career in ET, and I would encourage others to consider it for their academic careers.

Journey to Authenticity: Finding a Way for Women to Thrive in Engineering Technology

Elizabeth Betsy Dell

Background and Context

The need for diversity, mentoring, and inclusion in engineering has never been more critical. As technological challenges become increasingly complex and multifaceted, the importance of diverse teams in fostering innovation, improving performance, and ensuring representation and inclusion cannot be overstated. Despite women making up nearly half of the U.S. workforce, their representation in STEM fields, particularly in engineering technology, remains significantly low. This chapter explores the historical context and current landscape of diversity in engineering technology, highlighting the enduring gender disparities and the critical need for targeted efforts to support and retain women in these fields. This chapter serves as a call to action, urging stakeholders to prioritize diversity, mentoring, and inclusion in engineering technology. By doing so, we can unlock the full potential of diverse talent, drive innovation, and create a more just and equitable society.

E. B. Dell (✉)
Rochester Institute of Technology, Rochester, NY, USA
e-mail: emdmet@rit.edu

A. Lucietto (ed.), *Innovating for Tomorrow: Engineering Technology*, Women in Engineering and Science, https://doi.org/10.1007/978-3-032-07645-8_4

Historical Context and Importance of Diversity

Diversifying and growing the engineering workforce is needed to provide the diversity of thought required to find innovative solutions to ever-increasing and rapidly changing technological challenges (Hill et al., 2010). Attracting more women who are underrepresented in technical fields is one way to address this challenge. Despite women making up nearly half of the U.S. workforce, they represent only 27% of STEM workers and about 15% of engineering occupations (Martinez & Christnacht, 2021). Despite efforts to recruit and retain women in Science, Technology, Engineering, and Math (STEM) fields, women make up only 16% of the engineering workforce as of 2023 (SWE, 2024). Overall, women account for 24.8% of bachelor's degrees in engineering compared to 15.1% in engineering technology (ASEE, 2023). This underscores the need for targeted efforts to support and retain women in these fields. There is no clear reason for the lower representation of women in engineering technology (ET) programs. Engineering technology programs provide a more hands-on approach to educating future engineers with increased emphasis on application over theory. I have some thoughts as to why this may be. It could be that the "hands-on approach" may not appeal as strongly to women who may not have personal experiences with tools and machinery. Another theory is that women are more likely to be accepted into "traditional" engineering programs by meeting the math and science requirements that are needed. My experience with students indicates that many were simply not aware of "engineering technology" when researching colleges.

Not only do more women and other underrepresented minorities need to be enrolled in Engineering Technology programs, but they also need to be retained and graduate. Research indicates factors that contribute to low retention rates for undergraduate women in engineering programs include low academic self-confidence, feelings of isolation, a lack of established women engineers and faculty as role models, an unwelcoming academic climate, and demotivating and ineffective instructional approaches (Cadaret et al., 2017; Cordova-Wentling & Camacho, 2009; Dasgupta, 2011). Self-efficacy is a belief in one's own ability to be successful at a given task (Bandura, 1997). Self-efficacy has been demonstrated to correlate with the persistence of women in engineering disciplines (Falk et al., 2016; Marra et al., 2009). The self-confidence gained through a support network is a critical factor in degree completion for women (Goodman Research Group, 2002). Women value community and collaboration more than their male counterparts (Chesler & Chesler, 2002). A sense of belonging positively impacts academic achievement and retention in college. This is especially true for students from underrepresented groups in their major (Strayhorn, 2012). Peer support and a sense of belonging increase the persistence of women in STEM major (Rainey et al., 2018). Proactive steps such as providing a "women in [STEM major]" program and offering seminars and workshops can play an important role in integrating women into STEM departments (Hill et al., 2010).

The business case for more diversity in engineering is clear. Diverse teams bring a variety of perspectives and ideas, which can lead to more innovative solutions. Different backgrounds and experiences can inspire creative approaches to problem-solving (Lee et al., 2024). Studies have shown that diverse teams often perform better. They can draw on a wider range of skills and knowledge, leading to more effective decision-making and better outcomes (Hess et al., 2024; Page, 2008). A diverse engineering workforce ensures that the needs and perspectives of different communities are considered in the design and implementation of technology. This leads to more inclusive and user-friendly products and services (Riley, 2008). Additionally, encouraging diversity in engineering helps to address skill shortages by tapping into a broader talent pool. This is particularly important in STEM fields where there is a high demand for skilled professionals (Hunt et al., 2015). Companies that prioritize diversity often see economic benefits, including increased profitability and market share. A diverse workforce can better understand and cater to a diverse customer base (Herring, 2009). Promoting diversity in engineering is a matter of social justice. It ensures that everyone, regardless of their background, has the opportunity to contribute to and benefit from the advancements in technology (Bix, 2004).

Background: Personal Experiences and Stories

Early Interests In high school, I excelled in math and science, which led my guidance counselors to encourage me to explore engineering as a career option. My older sister was an engineer, which also influenced my decision. Initially, I did not fully understand what an engineer was. I enjoyed drawing and being creative, and I thought engineers designed the visual aspects of cars. It was not until I started engineering school that I realized this type of work was more aligned with industrial designers and artists.

Education When I began college, I struggled. High school had been easy for me, and I lacked good study skills. This made me question whether engineering was the right path. I even considered working in the skilled trades because I enjoyed hands-on work. However, as I adapted to college coursework, my grades improved, and I began to feel more confident in my abilities. Despite this, I found the focus on automotive engineering examples unappealing, as I did not have much knowledge about cars and their systems. There was an assumption that all engineering students, especially at a school with a strong automotive background, already knew these things. My perspective changed when I took a Polymer Science class. The course focused on the applications and science of plastics in engineering, and it sparked my interest. I saw how plastics could save energy, save lives, and be used innovatively and creatively. With a focus on application, I could see how my education could be utilized in my career. I pursued internships in materials laboratories, where I enjoyed the hands-on nature of materials testing and solving related problems. There was no option for “Engineering Technology” at my university, and I did not know about such programs. Such a program would have been of interest due to the hands-on nature or the education and the focus on application.

I attended what is now Kettering University for my Bachelor’s degree in Mechanical Engineering. The university was formerly known as General Motors Institute and GMI Engineering & Management Institute. It had a 5-year cooperative (co-op) education program where students alternated every 3 months between classes and co-op assignments. We were required to co-op with the same employer and complete a 9-month-long project in our fifth year, which included writing a thesis on an engineering project from our co-op employer. As a woman studying engineering, I faced several challenges that are typical for women in this field. Imposter syndrome and a sense of not belonging were constant companions. I was often afraid to speak up in classes full of men, fearing I might get an answer wrong. Many of my male peers had more experience with the tools we used in the lab, which made me hesitant to try and potentially fail in front of them. I did not enjoy many of the labs for this reason. At the same time, I did not want to be treated differently because I was a woman. I chose not to join the Society of Women Engineers and instead joined the American Society of Mechanical Engineers. At the time, I did not see the benefits of a network of women, partly because I already had a support system. I lived in a house with seven other women studying engineering. We were a close-knit group that naturally studied together and supported each other. We could be ourselves within our group, which was a stark contrast to the classroom setting.

During my internships in manufacturing plants, I encountered additional challenges. It was common to receive catcalls when walking on the plant floor, and I even had a colleague suggest I sit on his lap. In the maintenance area, where I spent a week learning about various trades, the all-male group had pornographic images openly displayed at their workstations, making it very uncomfortable for a young woman.

I also experienced the hierarchy and privilege that can occur in the workplace. As an intern, I could get items from the tool crib because I was considered an engineer, while the tradesmen needed a supervisor’s approval. This bothered them, and I understood why. Here I was, a 20-year-old with just a couple of years of school under my belt, trusted to know what items to get from the tool crib, while they, with years of experience, were not. This was one of my first exposures to the workplace hierarchy and privilege. I have always respected every person for what they bring to the table, not thinking I am any better or more knowledgeable simply because of a degree.

My co-op employer was Rochester Products, a division of General Motors, which changed names several times during my 7-year employment with the company. While my degree was not in Engineering Technology, I found myself drawn to more hands-on laboratory experiences, enjoying these co-op assignments the most. I decided to do my senior thesis on a project in the Materials Laboratory at AC Rochester. My project involved evaluating glass-filled nylon 6,6 for use in an intake manifold application. At the time, there were concerns about using a polymeric material in such a critical engine application. Proving the material could withstand the underhood environment was critical to its successful launch. The project also involved developing a technique using Fourier transform infrared spectroscopy to monitor polymer degradation, allowing for small samples to be evaluated from parts in use without removing the component. This technique was effective, and the results were published in a journal paper (Murty Dell & Yehl, 1990).

Career Path

It was during my time at GMI that I first considered becoming a professor. Our landlord, a professor at the university, would occasionally stop by and chat with us. I had been interested in possibly becoming a teacher, so the idea of combining engineering and teaching appealed to me.

Upon graduation, I landed a full-time position in the materials laboratory where I completed my thesis, working in a materials development role. I enjoyed the hands-on testing in a lab setting and solving problems related to performance and competitive analysis. After working in the materials lab for a few years, I transferred to a Design Engineering position, designing air induction systems. This involved a move from Rochester, NY, to Flint, MI. I was excited to be closer to many of my college friends. One experience I had while working in Michigan was a conference where we took and analyzed our Meyers-Briggs personality indicator assessment. My results were ENFJ: Extraverted, Intuitive, Feeling, and Judging. The types of careers that were described as appealing to those with this personality type were teaching, coaching, and counseling. The person who conducted the assessment pulled me aside and noted that my personality type was not the typical engineer, so he wondered why I decided to pursue engineering. Most of the engineers in the room were indeed of a different personality type: ISTJ—Introverted, Observant, Thinking, and Judging. It made me reflect on why I chose my career journey. While this experience was not the primary factor, I soon decided to quit my job and pursue graduate school full-time to become a professor. I enrolled in a PhD program in Macromolecular Science & Engineering at the University of Michigan. I sought a dissertation project that involved hands-on testing and found one studying the use of polymers for automotive fuel tanks and their permeation resistance to hydrocarbon fuels. I secured a summer internship at the General Motors Research Center during my graduate studies.

However, my research stalled due to differing views from my two advisors—one preferred a computational approach, while the other favored a lab-based science approach. I also realized that the career path of professors at the University of Michigan, a research-intensive university, was not what I wanted. I aspired to be primarily a teacher, but these professors, being at a research-intensive university, were primarily researchers and taught very few courses. Additionally, I was one of only two women in the program and the only woman who was not from a foreign country. Most of the graduate students were foreign students who developed their own study groups and support networks, making it difficult for me to find a similar support system. I recall studying for the qualifying exams with these students and noticing they had access to old exams and homework from almost every professor, which they used to succeed in various classes. This experience demonstrated how people naturally form networks with those similar to themselves, while "others" have a more difficult time forming networks and lose access to the knowledge and support these groups have.

When I was debating whether or not to continue my doctoral studies, I decided to return to the engineering workforce and continue my studies part-time. I secured a position at the Toyota Technical Center working as a Materials Development Engineer. The position was very hands-on with evaluating various materials for interior trim and seating applications. I had two distinct experiences while there that were gender-related. I realized whenever the department office assistant was out on vacation, the department supervisor would ask for me to assist with administrative tasks like sending faxes and mailing packages. When he first asked me, I thought it was because he did not know how. But then I realized there was a pattern. He was asking me and not my male colleagues. I did bring this to his attention. It was a learning experience in that when similar things happened in the future, I would refuse or at least question such requests immediately. I also vividly recall that when I returned from my honeymoon, one of my co-workers from Japan told me that in Japan, women do not return to work after they get married. One valuable experience I had during my time at Toyota was with working with engineers and supervisors from another country. Toyota was mindful of the need for awareness of the challenges and value of a multicultural workforce. They held team building sessions and workshops where there were differences in communication styles in Japan versus the U.S.

After leaving the graduate school with a Master's Degree, I soon had my first child. I spent 10 years as a stay-at-home mother. During this time, I started teaching as an adjunct at a local community college. I found teaching to be a role that really resonated with me. I was able to combine my interests in science with helping others. I recall it was not until my third or fourth semester of teaching that I realized after coming back from a break that this was actually a job and not just a hobby. I enjoyed teaching that much. I also engaged in volunteer and community efforts. I participated in and led science and math programs at the schools my children attended and served on the leadership board for a preschool and a group for stay-at-home mothers. One hobby I engaged in that would later assist me in my role as a professor and researcher was entering cooking contests. I won over $30,000 in cash and prizes and had recipes published in national publications of Cooking Light and Southern Living. When you submit to these contests, you have to carefully understand what the sponsor is looking for and explicitly follow the rules. When I began writing grant proposals as a faculty member, I found I had to use a very similar process: know what the sponsor is looking for, use their keywords, give clear explanations, and follow the rules.

After my children were in school, we moved away from where I had been teaching at a community college I began to look for teaching opportunities. The Rochester

Institute of Technology (RIT) had an opening for a faculty position in Packaging Science that did not require a PhD. I applied and made it to the on-site interviews but was not selected in the end. This role also included running a research center, and I did not have experience doing so. My application caught the eye of the department chair of the Manufacturing & Mechanical Engineering Technology Department. The department needed adjuncts who could teach during the day. I started as an adjunct teaching a Materials Science laboratory. The next semester, a faculty needed to take a leave of absence, so I taught a full load of Statics laboratories and Polymer Engineering laboratory. During this semester, a posting was made for a full-time faculty in the department. At the time, a doctoral degree was not required, only preferred. The requirements were a Master's Degree and a minimum of 3 years of industrial experience. While I was not sure I was ready to return to work full-time, there was a strong feeling that, before long, a doctoral degree would be required for all tenure-track faculty positions at RIT. I applied and was selected. The job offer was made over the phone by the department chair. The very first words that came out of his mouth were "I want you to know this offer is only because you are a woman." I was taken aback. There was part of me that did not want to take a job offered in such a manner, but this was my dream job. One that I had thought would not be possible, since I had foregone completing my dissertation. The chair went on to question me as to whether I really wanted to work full-time and suggested I could spend my afternoons doing something like playing tennis. Yes, he really said that! Despite all of this, I accepted the job. Over the first few years, these comments were always in the back of mind. I was convinced I would prove him wrong. I got to know the faculty who had served on the search committee and inquired as to whether or not my gender was a determining factor in my selection. They said it was absolutely not the case. I was the top choice of the search committee. Fortunately, the chair only stayed in the role for a few more semesters. I had a department with very supportive colleagues.

Two weeks after I started as an adjunct at RIT, I was asked to serve on a panel of women engineers for a new group of women students in the engineering technology program. This connected me with the group of women faculty leading and involved with the Women in Technology (WIT) program in the college. While it was meant to support the retention of our women students through building networks and providing professional development opportunities, it was also a peer mentoring network for the women faculty who were involved. I quickly became involved with the leadership team. We took turns planning events. At the time, RIT was a teaching-focused university with low research expectations of faculty. But soon there was a major change in research expectations. The faculty involved no longer were able to commit as much time to WIT as it was not valued as an activity that would lead to promotion and tenure. I was asked to become the Endowed Chair of the college and lead WIT. I took this on and applied for grants that would bring in sponsored research funding to support WIT and leveraged these funds to further research into the recruitment and retention of women students in Engineering Technology. A major grant was an NSF Scholarship in Science, Engineering, Technology and Math (S-STEM). This not only provided scholarship funding to recruit more women students and programming enhancements to improve the retention of women in the programs, but it also provided funds to hire a part-time coordinator of WIT. This came to be a critical aspect of an office to support diverse students in the college. When the grant ended, the college saw the impact of having dedicated support staff for these efforts and committed hard funding for the role. Over a few years, the role was expanded to a full-time position with oversight of programs for underrepresented minority students, in addition to women. The work of WIT was recognized by the national Women in Engineering Pro-Advocates Network when we received "Program of the Year" in 2012. My involvement in WIT was an important aspect for my achieving tenure. I secured six grants that supported WIT activities and published over a dozen conference proceedings and journal papers related to research-based approaches to supporting women students in STEM. An example is Supporting Women in Engineering Technology Programs in the Journal of Engineering Technology (Dell, 2019).

Over time, I became more involved in institutional efforts to support gender equity. I held the position of Faculty Associate to the Provost for Women Faculty. This role advises upper administration on best practices for supporting women faculty and current challenges. This role led to me being invited to join RIT's NSF ADVANCE team. RIT received an NSF ADVANCE Institutional Transformation grant to promote gender equity for women faculty in STEM (Dell et al., 2016). Through these roles, I valued the connections I made with women faculty across the campus in varying disciplines and roles. Through these connections, I found mentors, research collaborators, and mentees.

Practical Insights and Recommendations

Mentoring and Opportunities

In the realm of mentoring, I believe having a support network is critical to success and managing challenges. In college, this support network was formed from the all-women's dorm hall where I was assigned. This group of women, in some form with some added and some removed, continued living together for the 5 years of college. We studied together, shared resources, and supported each other not only with the challenges of college coursework but also with navigating life as college students.

When I started working full-time, the lack of women in engineering really struck me. As a young woman in engineering, finding that social and career support network was more of a challenge. Even something as simple as deciding who to sit with in the cafeteria was difficult. Conversations at the men's table often focused on sports, cars, hunting, or fishing—topics in which I had little interest or knowledge. There were not many women scientists and engineers, and the few women there were mostly mothers whose conversations seemed to focus on the best brand of diapers or their upcoming Tupperware party. I really missed the social support I had in college, which also provided peer mentors and academic support.

I had several mentors throughout this early stage of my career, mostly men, simply due to the numbers. When I started my career in academia as a professor, I was quickly invited to become part of a group that supported women students in Engineering Technology. This group was led by four other women faculty, and I again had a support network. By knowing these women, I was introduced to other women in different colleges at the university, expanding my network. This network not only provided career mentoring but also important support in achieving work–life balance. In my home department, there were only two women in a department with 16 faculty. The numbers have improved over the years. We now have four women on the tenure track/tenured and two full-time lecturers. But other departments have struggled to attract women. For example, the Electrical Engineering Technology Department has two women, one tenured and one lecturer.

It can be important to have mentors who are not in your own power structure so you can candidly ask questions about things like annual evaluations, requirements for promotion, how to ask for career support, and navigating service assignments. Through my networks, I eventually landed a role leading the main office for gender equity in faculty roles at my university. As Faculty Associate to the Provost for Women Faculty, I often met one-on-one with junior faculty, many of whom lacked the faculty network that I was fortunate to have. When I later became the director of the AdvanceRIT office, I was very mindful of the role this office could play in assisting faculty with developing these networks both inside and outside of their colleges. Unfortunately, gender bias and microaggressions are still present in universities, particularly in STEM fields. Knowing you are not alone and having receptive ears for sharing challenges and strategizing solutions is very helpful.

Authenticity and Inclusion

I strongly believe that the ability to be your authentic self in your work environment is critical for thriving in your position and career. There was a time at my university when a student group supporting women in engineering was told they should not be doing craft projects out in the open where others could see. The concern was that doing crafts was not technical in nature and would make the women appear less technically savvy. However, as an engineer, I have seen that many engineers engage in crafts and are creative people—of all genders—musicians, woodworkers, knitters, painters, etc. Being able to be who you are and share joint activities doing something you enjoy should not be frowned upon. If a group of men went on a golf outing, which is as directly related to engineering as a craft project, would that be seen as not technical enough? When I led the women faculty initiatives, we held "Pro-craftination" sessions for faculty to take a break from grading or writing and spend time with other faculty while working on a craft. We made earrings and holiday cards for charities, talked, connected, and found mentors and supporters. I believe corporations and universities need to support affinity groups of all kinds. If you walk into a room and see no one who looks like you or has experiences akin to your own, it leads to feelings of isolation. When someone does not have a network of colleagues at various points in their career, they can miss out on unwritten knowledge about how to navigate one's career.

There was a new faculty member in my department who had an infant and was not sure what to do when her child care canceled last minute. She wondered about bringing the well-behaved baby to class but did not know what people would think. I encouraged her to do so. I told her that the students saw her as a whole person—in this case, a mother—was a good thing. I still recall seeing her in front of the class with her baby strapped to her in a front carrier with the baby wearing a tie.

An example at my university was the practice of offering course release to faculty who welcomed a child into their family through birth or adoption. Some colleges and deans regularly supported this, but in other colleges, faculty did not even think this was something they could request and negotiate for. We charged the faculty senate with making a policy called a "modification of duties" policy. Now, it is in policy that faculty get a course release, and it happens fairly across the university, not only when there is a supportive department chair or dean. This also speaks to the importance of having supportive practices in policy versus relying on practices that may differ from one leader to another.

When part of the AdvanceRIT team, a group of the leaders of the project participated in a leadership development series. It was a transformative experience. We learned about our leadership styles and discussed approaches to leading teams and driving organizational change. However, there was one tip we received from the leader that I found problematic. We were encouraged to read the sports page before we came to work on Monday so we could chat with our male colleagues about sports. What?? While I appreciate the importance of building interpersonal connections, this suggestion felt off-putting. I believe in being my authentic self in the workplace and expect others to do the same. Women do not have to behave like men to succeed. Success in the workplace is about leveraging one's unique strengths, skills, and perspectives and valuing others' unique strengths, skills, and perspectives.

Recommendations

To create a more inclusive and equitable field in engineering technology, it is essential to implement targeted strategies and initiatives. Here are some key recommendations:

- Supportive Policies and Practices:
  - Develop and Enforce Anti-Discrimination Policies: Ensure that clear policies against discrimination and harassment are in place and strictly enforced. This includes providing training for all employees on recognizing and addressing bias and microaggressions.
  - Flexible Work Arrangements: Offer flexible work schedules and remote work options to accommodate diverse needs, including those of working parents and caregivers.

- Mentoring and Support Networks:
  - Establish Mentoring Programs: Create formal mentoring programs that pair experienced professionals with newcomers. These programs should focus on career development, skill-building, and navigating workplace challenges.
  - Peer Support Groups: Encourage the formation of peer support groups where individuals can share experiences, provide mutual support, and build a sense of community.
- Inclusive Recruitment and Retention Strategies:
  - Diverse Hiring Practices: Implement hiring practices that actively seek to diversify the workforce. This includes using diverse hiring panels and ensuring job postings reach a wide audience.
  - Retention Initiatives: Develop initiatives aimed at retaining diverse talent, such as professional development opportunities, leadership training, and pathways for career advancement.
- Educational and Professional Development:
  - Inclusive Curriculum: Ensure that educational programs and curricula are inclusive and reflect diverse perspectives. This includes incorporating examples and case studies that resonate with a wide range of students.
  - Continuous Learning: Provide ongoing training and development opportunities focused on diversity, equity, and inclusion. Encourage all employees to participate in these programs.
- Affinity Groups and Resource Networks:
  - Support Affinity Groups: Establish and support affinity groups for underrepresented populations within the organization. These groups can provide a platform for networking, advocacy, and professional growth.
  - Resource Networks: Create resource networks that offer access to information, tools, and support services related to diversity and inclusion.
- Leadership Commitment and Accountability:
  - Leadership Buy-In: Ensure that organizational leaders are committed to diversity and inclusion. Leaders should actively promote and participate in diversity initiatives and set a positive example for the rest of the organization.
  - Accountability Measures: Implement measures to hold leaders and employees accountable for progress on diversity and inclusion goals. This can include regular reporting, performance evaluations, and recognition of achievements in this area.

By adopting these recommendations, organizations can create a more inclusive and supportive environment for women and other underrepresented groups in engineering technology. These efforts will not only enhance individual career success but also drive innovation and improve overall organizational performance.

Conclusion

This chapter has highlighted the critical role of diversity, mentoring, and inclusion in the field of engineering technology. Through a personal narrative, the chapter explored my journey into engineering technology, detailing the challenges faced and the importance of mentoring and support networks. These were my experiences. I am aware that I am privileged in many ways. I have white skin, I am able-bodied, and English is my first language. Women with multiple identities face additional and more complex challenges. This narrative underscored the necessity of being one's authentic self in the workplace and the impact of inclusive environments on professional success.

Practical insights and recommendations were provided for fostering diversity and inclusion in engineering technology. These include supportive policies, affinity groups, and mentoring programs that create a more inclusive and equitable field. The chapter also highlighted the importance of proactive steps such as providing "women in [STEM major]" programs and offering seminars and workshops to integrate women into STEM departments.

Promoting diversity, mentoring, and inclusion in engineering technology is a strategic imperative for innovation and economic success. By continuing to support and implement these initiatives, we can create a more diverse, inclusive, and thriving engineering workforce.

References

ASEE. Engineering & engineering technology; By the numbers (2023). Retrieved from https://ira.asee.org/wp-content/uploads/2024/12/Engineering-Engineering-Technology-By-the-Numbers-2023-1-combined.pdf

A. Bandura, *Self-Efficacy: The Exercise of Control* (W.H. Freeman, New York, 1997)

A.S. Bix, From "engineeresses" to "girl engineers" to "good engineers": A history of women's US engineering education. NWSA J., 27–49 (2004)

M.C. Cadaret, P.J. Hartung, L.M. Subich, I.K. Weigold, Stereotype threat as a barrier to women entering engineering careers. J. Vocat. Behav. **99**, 40–51 (2017). https://doi.org/10.1016/j.jvb.2016.12.002

N.C. Chesler, M.A. Chesler, Gender-informed mentoring strategies for women engineering scholars: On establishing a caring community. J. Eng. Educ. **91**(Compendex), 49–55 (2002)

R. M. Cordova-Wentling, C. Camacho. University and personal factors that hinder and assist women when completing a degree in engineering. Paper presented at the 2009 ASEE Annual Conference and Exposition, June 14, 2009–June 17, 2009, Austin, TX,USA (2009)

N. Dasgupta, Ingroup experts and peers as social vaccines who inoculate the self-concept: The stereotype inoculation model. Psychol. Inq. **22**(4), 231–246 (2011). https://doi.org/10.1080/1047840X.2011.607313

E. Dell, Supporting women in engineering technology programs. J. Eng. Technol. (Fall) (2019)

E. Dell, M. Bailey, L. Clayton, J. Christman, R. Garrick, C. Marchetti, S. Mason, M. Valentine, Y. Verhoeven, Women in Engineering Technology at RIT: Strategies to Recruit, Retain and Advance Women Engineering Technology Faculty and Students; Conference for Industry & Education Collaboration; Austin, TX (February, 2016)

N.A. Falk, P.J. Rottinghaus, T.N. Casanova, F.H. Borgen, N.E. Betz, Expanding women's participation in STEM: Insights from parallel measures of self-efficacy and interests. J. Career Assess. **25**(4), 571–584 (2016). https://doi.org/10.1177/1069072716665822

Goodman Research Group, Final report of the women's experiences in college engineering (WECE) project, in *Funded as "A Comprehensive Evaluation of Women in Engineering Programs"*, (National Science Foundation Grant, Cambridge, MA, 2002) REC 9725521, Alfred P. Sloan foundation Grant 96-10-16

C. Herring, Does diversity pay?: Race, gender, and the business case for diversity. Am. Sociol. Rev. **74**(2), 208–224 (2009)

J.L. Hess, A. Lin, A. Whitehead, A. Katz, How do ethics and diversity, equity, and inclusion relate in engineering? A systematic review. J. Eng. Educ. **113**(1), 143–163 (2024). https://doi.org/10.1002/jee.20571

C. Hill, C. Corbett, A. Rose, *Why So Few* (American Association of University Women, Washington, D.C., 2010)

V. Hunt, D. Layton, S. Prince, Diversity matters. McKinsey & Company **1**(1), 15–29 (2015)

W.C. Lee, J. London, C.D. Hawkins Ash, C.M. Pee, M. Josiam, Broadening participation in engineering as a sociopolitical phenomenon: A systems perspective. Studies in Engineering Education **5**, 98 (2024). https://doi.org/10.21061/see.137

R.M. Marra, K.A. Rodgers, D. Shen, B. Bogue, Women engineering students and self-efficacy: A multi-year, multi-institution study of women engineering student self-efficacy. J. Eng. Educ. **98**(1), 27 (2009)

A. Martinez, C. Christnacht, Women making gains in STEM occupations but still underrepresented (2021). Retrieved from https://www.census.gov/library/stories/2021/01/women-making-gains-in-stem-occupations-but-still-underrepresented.html

E. Murty Dell, T. Yehl, Adaptation of photoacoustic Fourier transform infrared spectroscopy for studying the thermal oxidation of nylon 66 at 150°C correlated to mechanical properties. Polym. Eng. Sci. **30**(24), 1595–1598 (1990)

S. Page, *The Difference: How the Power of Diversity Creates Better Groups, Firms, Schools, and Societies-New Edition* (Princeton University Press, 2008)

K. Rainey, M. Dancy, R. Mickelson, E. Stearns, S. Moller, Race and gender differences in how sense of belonging influences decisions to major in STEM. Int. J. STEM Educ. **5**(1), 10 (2018). https://doi.org/10.1186/s40594-018-0115-6

D. Riley, Engineering and social justice, in *Engineering and Social Justice*, (Springer, 2008), pp. 47–106

T.L. Strayhorn. Exploring the impact of Facebook and Myspace use on first-year students' sense of belonging and persistence decisions. Journal of College Student Development **53**(6), 783–796 (2012)

SWE. Research: Employment of Women in Engineering (2024). Retrieved from https://swe.org/research/2024/employment/

Welded Paths: Forging a Life in Engineering Technology

Angie Hill Price

This chapter chronicles the journey of Dr. Angie Hill Price, a pioneering leader in engineering technology whose career exemplifies the transformative power of mentorship, resilience, and inclusive leadership. From her early struggles as an underprepared undergraduate to her current role as Associate Dean at Texas A&M University, Dr. Hill Price's story offers a compelling blueprint for women navigating the intersections of technical expertise, academic leadership, and personal fulfillment. Her narrative challenges traditional STEM success models, highlighting the value of service, relational leadership, and work–life integration. As part of a broader volume on women in engineering technology, this chapter contributes a vital perspective on how women are not only entering the field but also reshaping it from the inside out.

Leadership Forged in Experience

In engineering technology, leadership often emerges not from linear trajectories but from lives shaped by resilience, mentorship, and a deep commitment to service. Dr. Angie Hill Price, Associate Dean for Undergraduate Programs at Texas A&M University, exemplifies this path. Her story is one of transformation—from a

A. H. Price (✉)
Texas A& M University, College Station, TX, USA
e-mail: price@tamu.edu

A. Lucietto (ed.), *Innovating for Tomorrow: Engineering Technology*, Women in Engineering and Science, https://doi.org/10.1007/978-3-032-07645-8_5

struggling undergraduate to a respected academic leader, offering a compelling model for women navigating engineering and administration.

From Uncertainty to Clarity: Discovering Engineering Technology

Dr. Angie Hill Price's journey into engineering technology began not with certainty but with confusion and struggle, a reality that resonates with many women entering male-dominated STEM fields. Raised in a small Texas town with limited access to advanced coursework, she entered Texas A&M University as a mechanical engineering major, only to find herself overwhelmed and underprepared. The math was daunting, the content abstract, and the sense of belonging elusive.

She describes those early semesters as a search for "the good stuff" the hands-on, applied work she had imagined engineering to be. But instead of inspiration, she encountered frustration. It was not until she began studying alongside her then-boyfriend (now husband), who had switched to engineering technology, that she glimpsed a different kind of engineering—one that aligned with her upbringing in a family-run manufacturing business. His coursework in welding, design, and manufacturing felt familiar and exciting. It was, in her words, "what I thought engineering was supposed to be."

This moment of recognition was transformative. It reframed her understanding of engineering and revealed a path that honored her lived experience and practical intelligence. Switching to engineering technology was not just a significant change but a reclamation of identity. Within a year, she went from academic probation to academic distinction, earning a 3.67 GPA in her first semester, earning recognition on the Dean's honor roll, and rediscovering her confidence.

This clarity did not come from a textbook or a lecture—it came from exposure, mentorship, and the courage to pivot. For many women in engineering, especially those from underrepresented or nontraditional backgrounds, this clarity is often delayed by institutional blind spots and narrow definitions of success. Dr. Hill Price's story underscores the importance of visibility and access: had she not seen her partner's coursework, or had faculty not supported her transition, her talents might have remained untapped.

Her experience also highlights the value of engineering technology as a discipline that bridges theory and practice. It offers a more inclusive entry point for students who may not see themselves reflected in traditional engineering narratives. For Dr. Hill Price, it was the beginning of a lifelong commitment to making engineering education more accessible, more human, and more reflective of the diverse ways people learn and lead.

Mentorship and the Making of a Leader

If clarity gave Dr. Angie Hill Price direction, mentorship gave her momentum. Her transformation from a struggling student to a confident scholar was not a solitary journey—it was scaffolded by faculty who saw her potential and invested in her success. This kind of mentorship can be life-changing in a field where women are often underrepresented and underserved.

Dr. Hill Price's mentors did not wait for her to ask for help. They anticipated her needs, offered guidance before she knew to seek it, and created a culture of proactive support. This was especially critical in her early days in engineering technology, where she was often the only woman in the room. One of her most influential mentors, Dr. Daryle Morgan, encouraged her academic growth and planted the seed of graduate school—a path she had never imagined for herself.

His belief in her was more than motivational; it was catalytic. When she excelled in his class, he did not just acknowledge her performance—he celebrated it publicly, helping to dismantle the quiet doubts she carried. His mentorship extended beyond the classroom, guiding her to pursue a master's degree and eventually a PhD. He reminded her, especially during the most challenging moments of her doctoral journey, that "no one can take that degree away from you."

This kind of mentorship rooted in advocacy, visibility, and genuine care is especially vital for women in engineering technology. It challenges the isolation that many experience and replaces it with a sense of belonging and purpose. Dr. Hill Price's mentors did not just help her succeed; they helped her see herself as someone who belonged in engineering, academia, and eventually, leadership.

Dr. Hill Price carried this mentorship model forward as she transitioned from student to faculty member. She became a mentor to women students seeking guidance in a male-dominated field and male students who found her approachable and empathetic. Her office became a space where students could ask questions they were afraid to voice elsewhere—about coursework, career paths, or even what to wear to an interview.

In her administrative roles, she has institutionalized this ethos of mentorship. She advocates for structures that support early-career faculty, especially those from underrepresented backgrounds, and she models leadership that listens first and leads with empathy. Her story illustrates that mentorship is not a one-time act but a lifelong practice that shapes not only careers but cultures.

Balancing Academia, Industry, and Family

For many women in engineering, the question is not whether they can succeed—it is how they can succeed while also honoring the full complexity of their lives. Dr. Angie Hill Price's career is a testament to the power of intentional decision-making, adaptability, and shared responsibility in navigating the intersections of professional ambition and personal fulfillment.

Her path through academia was never linear. After completing her undergraduate degree, she entered the industry, only to find herself drawn back to graduate school by the encouragement of a mentor. She pursued a master's in industrial engineering while teaching and raising a family. Later, during her PhD, she juggled a full-time teaching load, doctoral coursework, and parenting—an extraordinary balancing act that required discipline and a strong support system.

One of the most defining chapters of her journey came when her husband was offered a position in Mexico. Rather than seeing this as a disruption, Dr. Hill Price saw it as an opportunity. She negotiated a leave of absence from the university, continued her PhD research remotely, and worked part-time as a welding engineer in industry. This period enriched her technical expertise and deepened her understanding of global engineering practices and cross-cultural collaboration.

Her ability to integrate family and career was not accidental—it resulted from deliberate choices. She and her husband chose to live near extended family, creating a network of support that allowed them to share parenting responsibilities and remain present in their children's lives. They approached family life as a partnership, coordinating schedules and making sacrifices to ensure both could thrive professionally.

This intentionality extended to her professional choices as well. She selected roles that aligned with her values and allowed her to maintain a sense of balance. Even as she rose through academia, she remained grounded in the belief that leadership should not come at the expense of family or well-being.

As an administrator, Dr. Hill Price brings this philosophy into her leadership. She fosters a culture of trust and flexibility, encouraging her team to prioritize their families and personal lives without guilt. She models what it means to lead with empathy, showing that being both ambitious and human is possible.

Her story challenges the myth that women must choose between career and family. Instead, it offers a blueprint for integrating the two through intentional planning, mutual support, and institutional advocacy. In doing so, Dr. Hill Price redefines what success looks like in engineering technology and expands the possibilities for those who follow in her footsteps.

Leadership Through Service and Connection

For Dr. Angie Hill Price, leadership was never about chasing titles—it was about showing up, doing the work, and building relationships that matter. Her rise to Associate Dean for Undergraduate Programs at Texas A&M University was not the result of a carefully plotted career ladder but rather the natural outcome of decades of service, trust, and community-building.

From the earliest stages of her academic career, Dr. Hill Price immersed herself in service. She joined committees, volunteered for initiatives, and took on roles that many might have avoided. But for her, service was not a burden but a strategy. It was how she learned the inner workings of the university, built a network of allies, and made herself visible in spaces where decisions were made.

This approach is especially significant for women in engineering technology, where leadership pathways are often less visible and less accessible. By engaging deeply in service, Dr. Hill Price created her visibility. She became known for her technical expertise, reliability, collaborative spirit, and ability to get things done.

Her leadership style is rooted in connection. She believes in knowing people, not just their titles but their stories, strengths, and struggles. This relational approach has made her an effective advocate for students, faculty, and staff. When she stepped into the role of Associate Dean, she already knew most of the people in the room. She had worked with them, supported them, and earned their trust.

One of the most potent aspects of Dr. Hill Price's leadership is her ability to translate service into systemic change. She does not just serve on committees; she uses those platforms to advocate for inclusive policies, mentor emerging leaders, and ensure that engineering education reflects the diversity of its students. Her time as Faculty Senate Speaker—three different terms—gave her a unique vantage point on institutional governance, and she has used that insight to shape more equitable academic structures.

She also understands that leadership is not about doing everything alone. When she was asked to step into the Associate Dean role, her predecessor mentored her and spent an entire summer coaching her. That experience reinforced her belief in succession planning, mentorship, and the importance of preparing others to lead.

Dr. Hill Price's story challenges traditional notions of leadership in engineering. It shows that service is not a detour from leadership—it is the foundation of it. It shows that connection is not a soft skill but a strategic asset. It also shows that women in engineering technology can lead not by conforming to existing models but by creating new ones rooted in empathy, collaboration, and purpose.

A Woman's Perspective in Engineering Technology

Dr. Angie Hill Price's experience as a woman in engineering technology offers a powerful lens into the subtle, often unspoken dynamics of gender in technical fields. For the first portion of her career, she was the only woman in her program. This reality shaped not only how she was perceived but also how she perceived herself and her role within the academic and professional community.

Unlike traditional engineering, which often emphasizes theoretical abstraction, engineering technology is grounded in application, systems thinking, and human-centered problem-solving. These qualities align closely with many women's strengths: collaboration, communication, empathy, and a holistic view of systems and people. Dr. Hill Price's career exemplifies how these attributes are not only valuable—they are essential.

She recalls how female students often approach her privately, expressing doubts about their abilities or feeling isolated in classrooms dominated by confident male peers. Many of these students assumed they were the only ones struggling. Dr. Hill Price, who had once felt the same way, reassured them with data and empathy: "I grade their papers," she would say. "I know others are struggling too—they're better at hiding it."

This insight—born of lived experience—allowed her to create a more inclusive and honest learning environment. She normalized struggle as part of the learning process and encouraged students to seek help, ask questions, and support one another. Her presence alone served as a quiet but powerful representation: a woman who had survived and thrived in engineering technology.

But visibility came with a cost. As the only woman, she was often asked to serve on numerous committees, represent diversity in various forums, and mentor a disproportionate number of students. While this "service tax" is a common experience for women in academia, Dr. Hill Price chose to view it as an opportunity. She leveraged these roles to build influence, expand her network, and advocate for systemic change.

Her perspective as a woman also informed her leadership style. She led with empathy, listened deeply, and created space for others to be heard. She understood that leadership in engineering technology is not just about managing systems—it is about understanding people. Whether mentoring a student, guiding a colleague, or shaping curriculum, she brought a relational intelligence that enriched every space she entered.

Dr. Hill Price's story illustrates that women in engineering technology do not just bring diversity—they bring depth. They challenge assumptions, broaden definitions of success, and model new leadership methods. Her journey is a reminder that the field is not only stronger with women in it but also that it is incomplete without them.

Work–Life Balance and Institutional Culture

For Dr. Angie Hill Price, work–life balance has never been a luxury—it has been a necessity, a strategy, and a value she has carried into every stage of her career. As a woman in engineering technology, she has navigated the dual demands of professional excellence and personal responsibility with intention and grace. Her story offers a compelling counter-narrative to the myth that success in STEM requires sacrificing one's personal life.

From the beginning, Dr. Hill Price and her husband approached their careers as a team. They made deliberate choices, like living near extended family and choosing jobs that allowed flexibility to ensure they could raise their children in a supportive environment. When one of their children was sick, they coordinated schedules. When her husband's job took them to Mexico, they adapted. She continued her PhD research remotely, worked part-time in industry, and cared for their young children while maintaining her academic trajectory.

This kind of balance was not always easy, but it was made possible by a culture of trust and flexibility at home and work. At Texas A&M, Dr. Hill Price found a department that valued family and understood that life does not pause for professional milestones. Her colleagues welcomed her children into the office when needed, and her supervisors supported her through significant life events, including her mother's stroke just weeks after she became Associate Dean.

Now in a leadership role, Dr. Hill Price is committed to paying that culture forward. She leads with empathy, encourages her team to prioritize their families, and models what it means to be present and productive. She understands that faculty and staff cannot do their best work if they are overwhelmed or unsupported. Her leadership is grounded in the belief that people are not just employees but individuals with whole lives, families, and needs that deserve respect.

This philosophy is essential for women in engineering technology, who often face heightened expectations and scrutiny. Dr. Hill Price's example shows that leading with compassion without compromising on excellence is possible. She argues that building sustainable, inclusive institutions is the only way.

Her story also highlights the importance of institutional culture in shaping individual success. When universities create environments supporting work–life integration, they do not just retain talent but cultivate leaders. Dr. Hill Price's journey is a powerful reminder that when women are supported, they do not just survive in engineering technology—they lead, innovate, and transform the field.

Looking Forward: Women Leading the Future of Engineering Technology

As the landscape of engineering technology continues to evolve, so too does the role of women within it. Dr. Angie Hill Price envisions a future where women are not only present in the field but are shaping its direction, culture, and values. Her perspective is grounded in experience, but her vision is firmly focused on what is next.

In a world increasingly driven by automation, artificial intelligence, and rapid technological change, Dr. Hill Price believes that engineering technology's human element will become more important. The ability to communicate across disciplines, empathize with users, lead diverse teams, and think critically about the social impact of technology will be essential. These are areas where women, often socialized to be collaborative, intuitive, and systems-oriented, can lead with distinction.

But leadership does not happen in a vacuum. It requires access, encouragement, and opportunity. Dr. Hill Price emphasizes the importance of early exposure, of helping young women see themselves in engineering technology before they have been told they do not belong. She advocates for mentorship, visibility, and storytelling as tools to increase participation. "Sometimes," she says, "just giving someone the idea that they can do this is enough to change their life."

Dr. Hill Price sees promise in student-led organizations and peer networks, which can foster leadership development.

Her hope for the future is not just more women in engineering technology—it is more women leading it. Women who bring their full selves to the table redefine what leadership looks like, build cultures of care, innovation, and inclusion, and, like Dr. Hill Price, use their positions not just to succeed but to lift others as they rise.

In her own words, "Engineering technology is about making things work. But it's also about making systems work for people. And that's where women can lead—with technical skill, vision, empathy, and purpose."

Conclusion: A Blueprint for Inclusive Leadership

Dr. Angie Hill Price's story is not just a personal narrative—it is a redefinition of what leadership in engineering technology can look like. Her journey, marked by struggle, mentorship, adaptability, and service, challenges the traditional STEM success blueprint. Instead of following a straight path, she forged one that is deeply human, relational, and inclusive.

Her life's work demonstrates that engineering technology is not only about machines, systems, and processes—it is about people. It is about creating environments where students feel seen, faculty are supported, and leadership is exercised with empathy and integrity. Dr. Hill Price has shown that technical excellence and emotional intelligence are not mutually exclusive—they are mutually reinforcing.

For women in engineering technology, her story offers both inspiration and instruction. It affirms that there is no single way to lead and no perfect path to follow. What matters is showing up, staying curious, building community, and lifting others as you climb. It is about recognizing that your lived experience—your challenges, your insights, your voice—is not a detour from leadership; it is the foundation of it.

As the field continues to evolve, the need for diverse leadership has never been greater. Women like Dr. Hill Price are not just filling roles but transforming them. They are expanding the definition of what it means to be an engineer, an educator, and a leader. In doing so, they are building a future where engineering technology works not just for systems but for society.

Navigating the Path: Insights from an Engineering Technology Professional

Alisa Millar Henrie

Introduction

Engineering technology is a dynamic and multifaceted field that is crucial in advancing modern industry. It bridges the gap between theoretical engineering principles and practical applications, making it indispensable in developing and implementing innovative solutions across various sectors, including aerospace, defense, manufacturing, and more. Despite its significance, the field has historically been male-dominated, presenting unique challenges and opportunities for women pursuing careers in engineering technology.

This chapter delves into the educational journey, professional experiences, and insights of Alisa J Millar Henrie, PhD, a seasoned professional in engineering technology. Through her narrative, we explore the pivotal moments that shaped her career, the obstacles she overcame, and her contributions to the field. Alisa's story begins with a childhood fascination with engineering, sparked by her interest in space exploration and hands-on projects. Her journey is marked by resilience, adaptability, and a commitment to hands-on learning, which have been central to her success in both academia and industry.

Alisa's experiences highlight the importance of practical applications in engineering technology and the value of mentorship and support networks. Her mentors and passion for teaching and mentoring influenced her transition from industry to

A. M. Henrie (✉)
The University of Alabama in Huntsville, Huntsville, AL, USA
e-mail: alisa.henrie@uah.edu

A. Lucietto (ed.), *Innovating for Tomorrow: Engineering Technology*, Women in Engineering and Science, https://doi.org/10.1007/978-3-032-07645-8_6

academia. As the Department Chair of Engineering Technology at the University of Alabama in Huntsville (UAH), Alisa has been instrumental in developing a program that addresses the growing need for engineering graduates in the Huntsville area. The program offers both online and on-campus courses, providing flexibility for students with diverse backgrounds and commitments.

Throughout this chapter, we will explore Alisa's insights into the field of engineering technology, her approach to teaching and mentoring, and her advocacy for diversity and inclusion in STEM fields. Her story serves as both an inspiration and a guide for aspiring engineers, offering practical advice and encouragement for navigating the path to success in engineering technology.

Early Inspiration and Educational Journey

Alisa Henrie's fascination with engineering began at a young age, rooted in her childhood experiences and innate curiosity. At the age of 11, Alisa's interest in engineering was sparked by her dream of becoming an astronaut. She vividly recalls making a list with two columns, one column listing her talents and one column listing her interests. She was determined to make good use of the talents granted to her by God in a career aligned with both her skills and passions. After carefully reviewing the list, astronaut was the career containing the most selections in both columns. She spent the rest of that summer researching the biographies and educational background of all current and previous astronauts. This research uncovered that most astronauts at the time had either a military pilot background or advanced degrees in mechanical engineering.

This early passion led her to explore various aspects of space and engineering, including building models of space shuttles and robots and creating a simulated space launch experience for her younger sister's 6th grade science class, complete with mission control and a life-size space shuttle cockpit made from refrigerator boxes. She began saving money from her summer jobs and eventually attended Space Camp in Huntsville, AL, which sparked her interest in the University of Alabama in Huntsville.

Alisa's father, a civil engineer, did not want her to pursue engineering education because it was male-dominated. Although he was very supportive of her education and helped her obtain each of her degrees, he never acknowledged her professional accomplishments or referred to her as an engineer. When she was a college freshman, they had a discussion where he was clear in his preference that she not be an engineer, but she was clear in her commitment to continue in the profession, and they never spoke about it again.

Growing up in rural Idaho, she was fortunate to have access to a private 2-year university in her hometown, which played a significant role in shaping her educational path. Despite her initial plans to attend a large university, Alisa chose to stay in her hometown and enroll in the local religious school, Ricks College (now BYU-Idaho). This decision proved to be a blessing both financially and academically. The faculty at the school were deeply committed to their roles, viewing their work as a calling rather than just a job. This environment fostered a supportive and nurturing atmosphere, allowing Alisa to thrive academically. She took summer courses, including Fortran programming, where she was one of only two students in the class. Her professor, Dr. Cartmill, played a crucial role in her early education, loaning her copies of textbooks and encouraging her to be his teaching assistant.

Alisa's educational journey continued as she transferred to Brigham Young University (BYU) after completing her associate's degree in general engineering. The transition to BYU marked a significant step in her academic development, exposing her to a larger and more diverse scholarly community. At BYU, Alisa formed strong study groups with friends from her previous school, providing camaraderie and mutual support. She pursued a master's degree in mechanical engineering, conducting research in compliant mechanism design with Dr. Larry Howell, a brilliant and supportive advisor. During this time, her hands-on approach to learning helped her develop several cost-effective kinematics lab experiences and create interdisciplinary connections across campus to share testing equipment in other colleges.

Throughout her educational journey, Alisa faced challenges that tested her determination and resilience. One notable challenge was her experience with calculus, which she initially struggled with due to the condensed 5-week summer course format and the emphasis on graphing calculators rather than foundational teaching. This led to several struggles during her undergraduate studies. Despite these difficulties, Alisa persevered, eventually mastering the subject through self-study and review. This experience taught her the importance of a strong foundational understanding and the value of persistence in overcoming academic obstacles.

Alisa's journey was also influenced by her mentors, who provided guidance and support at critical moments. Dr. Tracy Nelson, a technology professor, was particularly influential in her academic development. He encouraged students to engage in hands-on projects and directly apply theoretical knowledge to practical problems. This approach resonated with Alisa, who had always sought opportunities to work with equipment and see immediate results. Her experiences with mentors like Dr. Nelson reinforced her belief in the importance of practical applications and hands-on learning in engineering education.

Alisa Henrie's early inspiration and educational journey were characterized by a deep passion for engineering, a proactive approach to learning, and the support of dedicated mentors. Her experiences highlight the importance of hands-on learning, practical applications, and resilience in overcoming academic challenges. These foundational elements have been central to her success in both academia and industry, shaping her career and contributions to the field of engineering technology.

Professional Experiences and Challenges

After completing her master's degree, Alisa Henrie faced a significant choice: whether to continue her education with a PhD or enter the workforce. She chose to gain practical experience by joining Boeing, a decision that would profoundly shape her professional journey. At Boeing, Alisa was hired to be part of a team that would stretch the 747 to a full double-decker plane. However, the program had been canceled by the time the new graduates, including Alisa, arrived at Boeing.

With a large number of new engineers and no specific projects to work on, many of her colleagues left the company. However, Alisa's resilience and adaptability allowed her to thrive in this environment. She took advantage of the opportunity to learn from her supervisor, Frank Hashburger, one of the best design engineers she had ever worked with, and gained hands-on experience in manufacturing and aerospace. Frank provided technical guidance and taught Alisa important lessons about manufacturing, aerospace, and work culture. These lessons included the importance of effective communication, teamwork, and the ability to adapt to changing circumstances—skills crucial for success in any engineering career.

Alisa's dedication and hard work did not go unnoticed. She was offered a team lead position over a special project. This recognition of her capabilities and potential was a significant milestone in her career. However, Alisa had already decided to return to Brigham Young University (BYU) to support her husband in completing his bachelor's degree and to pursue her doctoral studies in mechanical engineering. This decision was driven by personal and professional considerations, highlighting Alisa's commitment to her family's well-being and long-term career goals. Returning to BYU, Alisa faced balancing her doctoral studies with family responsibilities.

During her doctoral studies, Alisa continued seeking hands-on learning opportunities and engaging research that aligned with her interests. She worked closely with one of her favorite professors, Dr. Brent L. Adams, to set up a scanning electron microscopy lab. This experience allowed Alisa to delve into materials research, further expanding her technical expertise and research skills. The microscopy lab was used by multiple groups across campus, and Alisa was able to train and be involved with a wide variety of projects. She enjoyed keeping the equipment running in good condition and has fond memories of phone conversations with the microscopy technician on a long-corded landline phone in the lab: a screwdriver in one hand, a microscope blueprints in the other, and the corded phone between the shoulder and neck, troubleshooting equipment malfunctions.

During her doctoral program, Alisa experienced several impactful personal events, including a second-trimester miscarriage that affected her health, the subsequent birth of her daughter, and the defense of her doctoral work while six-and-a-half months pregnant with her middle child. Balancing her growing family and her doctoral work would not have been possible without a support network. Several family members stepped in to assist. Two of her younger sisters attended BYU at the time and contributed childcare (in exchange for grocery money and access to a computer). Alisa's aunt (who had been her babysitter while her parents attended college) ran a daycare in the next town over and kindly accepted her daughter into her home several days a week. Alisa's husband also supported them as they both worked on advanced degrees for the same professor, Dr. Brent L Adams. Dr. Adams was also supportive, and many research meetings were held with the youngest Henrie running underneath the conference table. Intense academic and personal demands marked this period of her life, but Alisa's determination and resilience saw her through.

A move to New Mexico slowed down the completion of Alisa's doctoral degree but allowed her husband to accept employment after his graduation. During this time, Alisa taught as an adjunct at the University of New Mexico, where she helped develop an associate's degree specifically for lab technicians at Los Alamos National Laboratory. This degree was focused on helping technicians obtain a better understanding of their daily work and support career advancement. Between obtaining her doctoral degree and obtaining full-time employment, Alisa experienced a time she fondly refers to as her 10-year sabbatical.

A second move to Huntsville, AL (rocket city) brought Alisa back into the presence of aerospace engineering. Unpredictable health (eventually diagnosed as Lyme disease) and family needs kept Alisa from full-time employment at this time but uncovered other opportunities. She was able to work with several K-12 schools in the area and spearheaded "Space Week" for local elementary schools. This was an exciting time where Alisa was able to bring many educational opportunities to local students, including holding a moon rock, designing space launch containers, model rocket building, and visits from local NASA specialists. Alisa coordinated hands-on space-themed activities for 400–900 children each year from K-6, including designing space suits for marshmallow peeps.

Alisa's professional journey also included several years at a nonprofit organization called KTECH, affiliated with the nonprofit Kids to Love. KTECH primarily serves foster children aging out of foster care, helping them transition into living wage jobs and achieve financial security. The program also includes students from other underserved populations in the community and students eager to obtain an education and start a career without incurring substantial debt. At KTECH, Alisa taught mechatronics and robotics courses to help students obtain industry-recognized certificates such as the Siemens mechatronics technician certificate and Yaskawa robotics material handling certificate. These courses aim to equip students with the skills to secure stable employment and pursue further education.

Transition to Engineering Technology

While working at KTECH, Alisa learned about a new engineering technology program being developed at the University of Alabama in Huntsville (UAH). Initially reaching out to explore adjunct teaching opportunities, she discovered that the program director position was vacant. Alisa was hired temporarily just weeks before the program's launch in January 2020. This role required her to develop and teach courses from scratch, both on campus and online, a task she approached with dedication and enthusiasm.

The early days of the UAH engineering technology program were marked by intense effort and rapid development. Alisa developed 11 new college courses, teaching each before turning the course over to an adjunct faculty member. She was able to interview and hire adjunct faculty from local industry. This was made easier by Huntsville's rich engineering presence, which allowed for the hiring of engineers knowledgeable in current best practices. This period was a "crazy ride," as she described it, but it was also enriching.

The engineering technology program is designed to recognize and build on students' existing skills and experiences and addresses the growing need for engineering graduates in the Huntsville area. UAH supported an innovative approach when developing the program offering online and on-campus courses and accepting technical credits as electives. This flexibility enables students with associate degrees or military backgrounds to complete their bachelor's degrees while continuing to work.

One key aspect of the UAH engineering technology program is its focus on practical applications and hands-on learning. Alisa's approach to teaching is project-based, allowing students to present their work and demonstrate their understanding through practical applications. Project-based learning reinforces theoretical knowledge and develops critical thinking, problem-solving, and teamwork skills. Students learn to navigate the complexities of real-world projects, from initial concept to final implementation, gaining valuable experience that will serve them well in their careers.

Alisa continues to be involved in education and advocacy efforts including active membership in the Society of Women Engineers (SWE), the American Society for Engineering Education (ASEE), and the American Society of Mechanical Engineers (ASME). She also regularly teaches summer courses at UAH for middle-school science educators to better teach engineering through inexpensive hands-on activities through Tech-Trek, a program sponsored by the American Association of University Women (AAUW) and supported by UAH.

Insights and Perspectives

Alisa's experiences in engineering technology have provided her with valuable insights into the field. She finds that engineering technology often aligns more closely with students' initial perceptions of what engineering entails. Many students are drawn to the hands-on, practical aspects of the program, which involve working with equipment, engaging in team projects, and solving real-world problems.

Nontraditional students, including those with military backgrounds, working professionals, and those returning to education after a break, face unique challenges. Alisa's advocacy focuses on providing these students the support and resources to succeed. This includes flexible learning options, personalized mentoring, and programs designed to recognize and build on their existing skills and experiences. Building strong support networks is essential for students, particularly those from nontraditional backgrounds. These networks provide emotional support, academic guidance, and professional connections. Alisa's experiences highlight the importance of fostering a supportive and inclusive environment where students can thrive.

Alisa's commitment to empowering women in engineering is evident in her advocacy and mentoring efforts. She acknowledges the unique challenges faced by women in the field, such as balancing advanced education with family commitments and overcoming gender biases. Alisa strives to provide the support and resources needed to help women overcome these obstacles and succeed in their careers.

Alisa's commitment to her family is evident in her approach to work–life balance. She has three children, and her experiences as a mother have shaped her perspective on balancing career and family responsibilities. Alisa's ability to navigate these challenges serves as an inspiration to other women in the field.

Advocacy for nontraditional students and women in engineering remains a priority for Alisa. She continues to support and empower students in STEM careers, striving to foster a diverse and inclusive engineering community. Her dedication to teaching and mentoring is central to her future goals, and she finds great joy in helping students learn and succeed. Alisa's vision for the future includes becoming a better teacher and mentor. She is committed to continuously improving her teaching methods, seeking new ways to engage and inspire her students. Her goal is to create an environment where all students can thrive and contribute to innovative solutions in engineering technology.

Conclusion

Alisa Henrie's journey in engineering technology is marked by resilience, innovation, and a commitment to excellence. Her experiences highlight the importance of hands-on learning, practical applications, and teamwork in engineering. Through her narrative, we gain valuable insights into the challenges and opportunities in the field and the strategies for success. Alisa's dedication to her students and her advocacy for diversity and inclusion inspire aspiring engineers, offering practical advice and encouragement for navigating the path to success in engineering technology.

Part III
Faculty: All Levels of University and Community College

Engineering technology education is a dynamic and practice-oriented field that relies heavily on its faculty's expertise, adaptability, and dedication. From community colleges to research-intensive universities, faculty members are the architects of student learning, the stewards of curriculum development, and the essential link between academia and industry. This module, *Faculty: All Levels of University and Community College in Engineering Technology*, brings together a diverse group of educators whose experiences span the full spectrum of academic institutions and professional roles.

At the community college level, faculty are often the first point of contact for students entering the engineering technology pipeline. These educators play a critical role in shaping student identity, building technical confidence, and preparing learners for direct entry into the workforce or transfer to 4-year institutions. The chapters from community college faculty emphasize the importance of accessibility, hands-on instruction, and the integration of industry certifications and workforce-aligned competencies. They also explore the challenges of teaching in resource-constrained environments and the creative strategies used to maintain student engagement and success.

Faculty at 4-year institutions, including regional universities and polytechnic campuses, face different expectations. In addition to teaching, they are often involved in applied research, curriculum innovation, and industry collaboration. These educators must balance classroom responsibilities with service commitments, accreditation preparation, and mentoring both undergraduate and graduate students. Their chapters delve into the evolving nature of engineering technology curricula, integrating emerging technologies such as automation and sustainability and the importance of interdisciplinary teaching and project-based learning.

Senior faculty and those in leadership roles provide a broader institutional perspective. Their contributions examine the responsibilities of mentoring junior faculty, leading programmatic change, and navigating the complexities of academic governance. These authors reflect on the importance of faculty development, succession planning, and cultivating inclusive departmental cultures. They also address the role of faculty in accreditation processes, strategic planning, and maintaining alignment with industry standards and expectations.

Across all institutional types, the authors in this module share a deep commitment to student success, workforce readiness, and the continuous improvement of engineering technology education. Their stories underscore the importance of adaptability, collaboration, and lifelong learning in a constantly evolving field. Whether teaching introductory courses in a community college lab or leading a capstone design project at a university, engineering technology faculty are united by a common purpose: to prepare the next generation of technologists, innovators, and problem-solvers.

This part invites readers to explore the lived experiences of faculty at every level of the academic spectrum. It offers insights into the diverse pathways into teaching, the rewards and challenges of academic life, and the critical role faculty play in shaping the future of engineering technology education.

This part brings together the voices and experiences of eight leaders in the field: Dr. Lesley Berhan, Dr. Anne Lucietto, Dr. Brittney Newell, Prof. A. Fey Barber-Dansby, Prof. Beth McGinnis-Cavanaugh, Dr. Jérika Lamas, Dr. Upinder Kaur, and Dr. Devon Pessler. Each author offers a unique perspective on the evolving responsibilities of faculty, working their way through various industrial experiences, scholarships, and the development of their own research goals and achievements.

From Trinidad to Toledo: Lesley Berhan's Journey in Engineering Technology

Lesley Berhan

Introduction

Lesley Berhan's journey from a third-world country to a prominent researcher in engineering technology is a testament to resilience, adaptability, and a passion for education. Born and raised in Trinidad, Lesley's path was shaped by a unique educational system and a series of pivotal moments that guided her toward a career in engineering and academia. This chapter explores her academic journey, challenges, and impactful research on engineering technology students.

Lesley's story begins in Trinidad, a vibrant island nation known for its rich cultural heritage and diverse population. Trinidad's educational system, like the other islands in the English-speaking Caribbean, is inherited from the British, with a strong emphasis on academic excellence and early specialization. From a young age, Lesley demonstrated exceptional aptitude in the sciences and mathematics, which set her on a trajectory toward engineering. Her journey was marked by a series of critical decisions and opportunities that would ultimately lead her to the United States and a successful career in academia.

Growing up in Trinidad, Lesley was exposed to a rigorous educational framework that required students to choose their academic paths early on. At 11, she took an exam that determined her placement in secondary school, where she excelled in subjects like mathematics, physics, and chemistry. This early specialization allowed

L. Berhan (✉)
University of Toledo, Toledo, OH, USA
e-mail: noreply@box.com

A. Lucietto (ed.), *Innovating for Tomorrow: Engineering Technology*, Women in Engineering and Science, https://doi.org/10.1007/978-3-032-07645-8_7

her to develop a strong foundation in the sciences, paving the way for her future studies in engineering.

Lesley's academic prowess earned her multiple scholarships, enabling her to pursue higher education without financial burden. Her undergraduate studies in civil engineering at the University of the West Indies in Trinidad gave her a solid grounding in structural engineering, a field that intrigued her due to its practical applications and societal impact. However, Lesley's ambitions extended beyond her home country, and she aspired to further her education in the United States, a decision that would prove to be transformative.

A scholarship from Enron facilitated Lesley's transition to the United States, funding her master's degree at MIT. This opportunity allowed Lesley to immerse herself in a new educational environment characterized by cutting-edge research and technological advancements. Despite initial challenges, such as adapting to the technological demands of her coursework, Lesley completed her master's degree, demonstrating her resilience and adaptability.

Lesley's journey did not stop there. After completing her master's degree, she moved to the Metro Detroit area, where she worked as a structural engineer. However, she soon realized her true passion lay in academia and research. She decided to pursue a PhD in mechanical engineering at the University of Michigan, focusing on nanocomposites. This decision was driven by her desire to explore new areas of engineering and ensure that her career options remain broad.

Lesley's academic career began at the University of Toledo, where she secured a position as an assistant professor of mechanical engineering. After obtaining tenure and being promoted to an associate professor, Lesley's research interests evolved toward engineering education, focusing on equity, access, and the experiences of engineering and engineering technology students. Her role expanded to include administrative responsibilities, particularly in student success and diversity initiatives.

Her curiosity about the differences between engineering and engineering technology paths drives Lesley's research on engineering technology students. She explores the motivations behind students' choices and the demographic factors that influence these decisions. Her work aims to understand how situational and structural factors impact students' success and retention in engineering technology programs.

One of Lesley's most gratifying projects was developing a pre-major program at the University of Toledo. This initiative provides alternate pathways for students who may not meet traditional admission criteria but have the potential to succeed in engineering. The program has significantly improved retention and success rates among these students by offering additional support and resources.

Lesley Berhan's journey from Trinidad to Toledo is a remarkable example of how resilience, adaptability, and a passion for education can shape a successful career in engineering technology. Her research on engineering technology students highlights the importance of understanding diverse pathways and providing equitable access to education. Lesley's story inspires aspiring engineers and educators and demonstrates the transformative power of education and research.

Early Education in Trinidad

Lesley's educational journey began in Trinidad, a vibrant island nation with a unique and rigorous educational system. The education system in Trinidad is structured to emphasize academic excellence and early specialization, which significantly influences students' career paths. Unlike the United States, where students often have more flexibility in choosing their subjects and career trajectories, Trinidad's system requires students to make critical decisions based on their academic performance at a young age.

In Trinidad, students' paths are often dictated by their performance in standardized exams and the subjects they choose in high school. At the age of 11, Lesley took an exam that determined her placement in secondary school. This exam was a pivotal moment in her educational journey, as it set the stage for her future academic pursuits. Lesley's exceptional performance in this exam allowed her to attend a prestigious secondary school, where she continued to excel in her studies.

During her secondary school years, Lesley demonstrated a remarkable aptitude for sciences and mathematics. The curriculum in Trinidad's secondary schools is designed to provide students with a strong foundation in their chosen subjects, preparing them for higher education and specialized careers. Lesley's proficiency in mathematics and physics was evident, and she quickly became one of the top students in her class.

In Trinidad, secondary education is divided into two main stages: O and A levels. O levels, or Ordinary levels, are typically completed by students around 16. These exams cover a broad range of subjects, including mathematics, physics, chemistry, biology, and additional mathematics (calculus). Lesley's performance in her O levels was outstanding, earning her distinctions in multiple subjects and solidifying her reputation as a top student.

Following her success in O levels, Lesley pursued A levels, or Advanced levels, which are more specialized and focused on fewer subjects. A levels are typically completed by students around the age of 18 and are crucial for university admissions. Lesley decided to specialize in mathematics, further mathematics, and physics for her A levels, a decision that naturally led her to consider a career in engineering. Her choice of subjects was influenced by her strong interest in these areas and her desire to pursue a field that combined theoretical knowledge with practical applications.

Lesley's academic achievements in A levels were equally impressive. She excelled in both mathematics and physics, demonstrating a deep understanding of complex concepts and problem-solving skills. Her performance in these exams opened doors to numerous opportunities for higher education, both in Trinidad and abroad.

The rigorous educational system in Trinidad, combined with Lesley's exceptional academic performance, set her on a trajectory toward engineering. Her early specialization in sciences and mathematics provided her with a solid foundation for her future studies and career. Lesley's journey from secondary school to university was marked by a series of critical decisions and opportunities that would ultimately lead her to the United States and a successful career in academia.

Pursuing Higher Education

Lesley chose to pursue civil engineering for her undergraduate studies, driven by personal interest and the perceived need for civil engineers in Trinidad. The decision was influenced by several factors, including her strong academic background in mathematics and physics, her desire to engage in a field with practical applications, and the opportunities available in Trinidad's infrastructure development sector.

Civil engineering appealed to Lesley because it combined theoretical knowledge with tangible outcomes, allowing her to contribute to developing and improving her community's infrastructure. She enrolled in a civil engineering program at a local university, where she specialized in structural engineering. This specialization focused on designing, analyzing, and constructing structures such as buildings, bridges, and other infrastructure projects.

The British-style education system in Trinidad provided Lesley with a rigorous and comprehensive curriculum that emphasized depth in specific subjects. Unlike the American education system, which often includes a broad range of general education courses, the British system is more focused on specialized subjects from the outset. This approach allowed Lesley to dive deeply into civil engineering topics, gaining a thorough understanding of structural analysis, materials science, and construction techniques.

Throughout her undergraduate studies, Lesley excelled academically, earning top grades and distinctions in her courses. Her dedication to her studies and her passion for engineering were evident in her academic achievements. She also gained practical experience through internships and projects, working on real-world engineering challenges that further solidified her interest in the field.

After completing her degree in civil engineering, Lesley worked for a year in Trinidad, applying her knowledge and skills to various infrastructure projects. This experience provided her with valuable insights into the practical aspects of civil engineering and the challenges faced by engineers in the field. It also reinforced her desire to further her education and expand her expertise.

Lesley aspired to continue her studies in the United States, where she believed she could access advanced research opportunities and cutting-edge technologies. She was particularly interested in pursuing a master's degree at a prestigious institution that would allow her to deepen her knowledge and engage in innovative research. Her ambition was driven by a desire to contribute to the global engineering community and to bring back valuable knowledge and skills to Trinidad.

The opportunity to study in the United States came through a scholarship from Enron, which funded her master's degree at MIT. This scholarship was a turning point in Lesley's educational journey, providing her with the resources and support needed to pursue her dreams. At MIT, Lesley was exposed to a new educational environment that emphasized research, innovation, and technological advancement. Despite initial challenges, such as adapting to the technological demands of her coursework, Lesley successfully completed her master's degree in civil engineering.

Lesley's pursuit of higher education in the United States was a transformative experience that broadened her horizons and opened up new opportunities for her career. It allowed her to engage with leading experts in the field, participate in groundbreaking research, and develop a global perspective on engineering challenges. This experience laid the foundation for her future academic and professional achievements, ultimately leading her to a successful career in academia and research.

Transition to the United States

Lesley's transition to the United States was significantly facilitated by a scholarship from Enron, which funded her master's degree at the Massachusetts Institute of Technology (MIT). This opportunity was a pivotal moment in her educational journey, allowing her to immerse herself in a new and challenging educational environment. At MIT, Lesley was exposed to advanced engineering concepts and cutting-edge research, which broadened her academic horizons and deepened her understanding of civil engineering.

Despite the excitement of this new chapter, Lesley faced several initial challenges. One of the most significant hurdles was adapting to the technological demands of her coursework. Coming from Trinidad, where her education had not heavily emphasized computer skills, Lesley needed to quickly become proficient with various software and tools integral to her studies at MIT. For instance, she had to learn to use spreadsheets and other computer applications essential for her project management course. This was a steep learning curve, as she had limited prior experience with such technology. The challenge seemed so daunting at the time that she dropped the course.

However, Lesley's determination and resilience helped her overcome other obstacles. She dedicated herself to mastering the necessary technological skills, often seeking help and utilizing resources available at MIT. Her hard work paid off, and she completed her master's degree in civil engineering within three semesters. This achievement marked a significant academic milestone and demonstrated her ability to adapt and thrive in a demanding and unfamiliar environment.

Lesley's experience at MIT was transformative, equipping her with the knowledge and skills that would later shape her professional career. It also underscored the importance of adaptability and continuous learning, qualities that she carried forward into her subsequent endeavors.

Career and Academic Choices

After completing her master's degree at MIT, Lesley moved to the Metro Detroit area, where she began her professional career as a structural engineer at Albert Kahn Associates, a well-known structural engineering firm. In this role, she was involved in various projects, primarily focusing on retrofitting existing structures to accommodate new equipment or modifications. This work, while valuable, often involved repetitive tasks such as reinforcing trusses to support additional loads on rooftops, which Lesley found less stimulating over time.

During her tenure at Albert Kahn Associates, Lesley began to reflect on her career aspirations and realized that while she enjoyed the theoretical aspects of structural engineering, the practical, day-to-day tasks did not fully engage her interests. She found herself yearning for the intellectual stimulation and dynamic environment of academia, where she could explore new ideas and contribute to the advancement of engineering knowledge.

Driven by this realization, Lesley decided to pursue a PhD in mechanical engineering at the University of Michigan. This decision was influenced by several factors, including her desire to delve into new areas of engineering and to keep her career options open. She chose to focus her doctoral research on nanocomposites, a cutting-edge field that combines materials science and engineering to develop new materials with enhanced properties. This area of study not only aligned with her interests but also promised to open up diverse opportunities in both academia and industry.

Lesley's doctoral journey at the University of Michigan was marked by rigorous research and academic challenges. She immersed herself in the study of nanocomposites, investigating their potential applications and the underlying mechanisms that govern their behavior. Her research contributed to the growing body of knowledge in this field and positioned her as an expert in a highly specialized area of mechanical engineering.

Upon completing her PhD, Lesley accepted a faculty position at the University of Toledo, where she has since built a distinguished career as a professor of mechanical engineering. Her work continues to bridge the gap between research and education, and she remains dedicated to fostering the next generation of engineers.

Academic Career at the University of Toledo

Lesley's academic career began at the University of Toledo, where she secured a position as an assistant professor of mechanical engineering. This role marked the beginning of a significant and impactful chapter in her professional journey. Initially focused on teaching and research within the mechanical engineering department, Lesley's responsibilities post-tenure and promotion eventually expanded to include various administrative roles, reflecting her commitment to student success and diversity initiatives.

As a professor, Lesley was dedicated to providing high-quality education and mentorship to her students. She taught a range of courses in mechanical engineering, sharing her expertise in dynamics, mechanics of materials, and other advanced engineering topics. Her teaching philosophy emphasized not only the technical aspects of engineering but also the importance of critical thinking, problem-solving, and ethical considerations in the field.

Recognizing the importance of supporting students beyond the classroom, Lesley took on administrative responsibilities aimed at enhancing student success. She became actively involved in initiatives designed to improve retention rates, academic performance, and overall student satisfaction. Her efforts included developing programs that provided additional academic support, career counseling, and opportunities for professional development. Lesley's approach was holistic, addressing the diverse needs of students and ensuring they had the resources and guidance necessary to succeed.

In addition to her work on student success, Lesley was deeply committed to promoting diversity within the College of Engineering. She spearheaded several initiatives aimed at increasing the representation of underrepresented groups in engineering programs. This included outreach efforts to attract more women and minority students to the field, as well as creating a more inclusive and supportive environment for all students. Lesley's leadership in these areas was instrumental in fostering a culture of equity and inclusion within the college.

Lesley's research interests also evolved during her time at the University of Toledo. Initially focused on technical aspects of mechanical engineering, her research began to shift toward engineering education. She became particularly interested in issues of equity and access, exploring how different demographic groups experienced engineering education and identifying barriers that might prevent certain students from succeeding. Her research aimed to develop strategies to improve access to engineering education for all students, regardless of their background.

One of Lesley's key research areas was the experiences of engineering technology students. She investigated the differences between traditional engineering programs and engineering technology programs, examining how these differences impacted student outcomes. Her work sought to understand the unique challenges faced by engineering technology students and to develop interventions that could support their success. This research was driven by her belief that all students should have the opportunity to pursue their interests in engineering and technology and that educational institutions should provide pathways that accommodate diverse learning styles and career goals.

Lesley's contributions to the University of Toledo have been profound, shaping both the academic and administrative landscapes of the College of Engineering. Her dedication to student success, diversity, and equity has left a lasting impact, and her research continues to inform best practices in engineering education.

Research on Engineering Technology Students

Lesley's research on engineering technology students is driven by her curiosity about the differences between traditional engineering and engineering technology paths. This interest stems from her observations at the University of Toledo, where both engineering and engineering technology programs are housed within the same College of Engineering. This unique arrangement provides a rich environment for comparative studies and offers insights into how these two educational tracks interact and complement each other.

Engineering technology programs at the University of Toledo are designed to be more application-oriented, focusing on the practical aspects of engineering. These programs often attract students who are interested in hands-on learning and direct application of engineering principles in real-world scenarios. In contrast, traditional engineering programs tend to emphasize theoretical foundations and advanced mathematical concepts, preparing students for roles that may involve more design and research.

Lesley's research explores the motivations behind students' choices between these two paths. She investigates whether students actively choose engineering technology over traditional engineering based on their interests and career goals, or if other factors influence their decisions. These factors can include situational elements such as work commitments, financial considerations, and personal circumstances, as well as structural elements like the availability of part-time study options and the flexibility of the curriculum.

One of the key aspects of Lesley's research is understanding the demographic factors that influence these decisions. She examines how different groups of students, including those from underrepresented backgrounds, navigate their educational choices. Her work aims to identify any disparities in access to engineering education and to understand how these disparities impact student success and retention in engineering technology programs.

Lesley's research also delves into the experiences of engineering technology students once they are enrolled in their programs. She looks at how situational and structural factors, such as the need to balance work and study or the availability of support services, affect their academic performance and overall satisfaction. By identifying the challenges these students face, Lesley aims to develop strategies to improve their educational experiences and outcomes.

The unique relationship between engineering and engineering technology programs at the University of Toledo provides a valuable context for Lesley's research. The close proximity of these programs within the same college allows for a collaborative approach to addressing common issues and sharing best practices. This integration fosters a supportive environment where students can benefit from the strengths of both tracks and where faculty can work together to enhance the overall quality of engineering education.

Lesley's work is particularly impactful because it addresses the broader goals of equity and access in engineering education. By understanding the factors that influence students' choices and experiences, she aims to create more inclusive and supportive educational pathways. Her research contributes to the development of policies and programs that can help all students, regardless of their background, succeed in their chosen field of engineering or engineering technology.

Impactful Projects and Initiatives

One of Lesley's most gratifying projects is the development of a pre-major program at the University of Toledo. This initiative was born out of her recognition that many students possess the potential to succeed in engineering but may not meet the traditional admission criteria due to various factors such as educational background, socioeconomic status, or lack of access to preparatory resources. Lesley saw an opportunity to create a more inclusive pathway that would allow these students to pursue their engineering aspirations.

The pre-major program is designed to provide alternate pathways for students who demonstrate potential but require additional support to bridge the gap between their current academic preparation and the demands of an engineering curriculum.

The program also includes bridge courses specifically designed to prepare students for the rigors of an engineering curriculum. These courses cover essential topics and skills that students may need to strengthen before transitioning to their major. The bridge courses are taught by experienced faculty who understand the unique challenges faced by pre-major students. Recognizing the importance of a supportive community, the pre-major program fosters a sense of belonging among its participants. Students are encouraged to engage in extracurricular activities, join engineering clubs, and participate in events that promote collaboration and networking. Building a strong community helps students feel connected and supported, which is crucial for their retention and success.

The program includes regular monitoring and assessment to track students' progress and identify areas where additional support may be needed. This proactive approach ensures that students receive timely interventions and resources to address any academic or personal challenges they may encounter. The impact of the pre-major program has been significant. By offering additional support and resources, the program has substantially improved retention and success rates among students who might otherwise struggle to meet the demands of an engineering curriculum. Many students who enter the program with initial academic deficiencies go on to successfully transition into their chosen engineering majors and graduate with their degrees.

Lesley's work on this initiative has not only helped individual students achieve their academic goals but has also contributed to the broader goals of equity and access in engineering education. The pre-major program serves as a model for other institutions seeking to create inclusive pathways for students from diverse backgrounds. It demonstrates that with the right support and resources, all students can succeed in engineering, regardless of their starting point. This project is particularly meaningful to Lesley because it aligns with her commitment to fostering student success and promoting diversity within the field of engineering. It reflects her belief that every student deserves the opportunity to pursue their dreams and that educational institutions have the responsibility to provide the necessary support to make those dreams a reality.

Wrap-Up

Lesley Berhan's journey from Trinidad to Toledo is a testament to the transformative power of education and the importance of resilience and adaptability. Her experiences highlight the significant impact that dedicated educators and inclusive programs can have on student success. From her early days in Trinidad, excelling in the sciences, to her advanced studies at MIT and the University of Michigan, Lesley's path has been marked by a commitment to learning and growth.

At the University of Toledo, Lesley has not only distinguished herself as a professor of mechanical engineering but also as a leader in student success and diversity initiatives. Her research on engineering technology students underscores the need to understand diverse educational pathways and to provide equitable access to opportunities. The development of the pre-major program is a shining example of how targeted support and resources can help students overcome barriers and achieve their academic goals.

Lesley's story serves as an inspiration to aspiring engineers and educators, demonstrating that with determination, support, and a passion for education, it is possible to make a lasting impact. Her work continues to shape the field of engineering technology, fostering a more inclusive and supportive environment for all students. As we reflect on Lesley's journey, we are reminded of the importance of creating pathways that allow every student to succeed, regardless of their background or starting point.

Blueprints and Breakthroughs: Anne Lucietto's Path into Engineering Technology

Anne Lucietto

The journey of Anne Lucietto, PhD into engineering/engineering technology began not in a classroom but in her family's tool and die shop outside Chicago. As a child, she was endlessly curious—disassembling toys, appliances, and anything mechanical she could get her hands on. Her father, Ledo Lucietto, unlike many men of his generation, encouraged Anne's fascination with machines. While others questioned his support for a daughter pursuing engineering, Ledo stood firm, nurturing her interest in a field still largely dominated by men.

Anne's early exposure to mechanics and manufacturing laid the foundation for her future. She initially considered careers in medicine and music, but her heart was

A. Lucietto (✉)
Purdue University, West Lafayette, IN, USA
e-mail: lucietto@purdue.edu

A. Lucietto (ed.), *Innovating for Tomorrow: Engineering Technology*, Women in Engineering and Science, https://doi.org/10.1007/978-3-032-07645-8_8

drawn to engineering. She started in music and biology at the College of St. Teresa, ultimately earning her Bachelor of Science in Mechanical Engineering from Marquette University, where she focused on design, materials, and thermodynamics. Her career began in industry, where she held progressively responsible roles across mechanical, electrical, civil, chemical, and environmental engineering domains. She worked in energy generation, nuclear plant construction, chemical processing, and materials fabrication—gaining a reputation for her expertise in facilities, maintenance, construction, and manufacturing management, also earning an MBA along the way.

After nearly three decades in industry and part-time teaching at community colleges, Anne transitioned fully into academia. She pursued a PhD in Engineering Education at Purdue University, driven to better understand and support engineering technology students. Her research focused on sustainable energy and students' educational pathways in applied engineering.

Anne is now an Associate Professor at Purdue University's School of Engineering Technology, a respected educator and mentor. She is known for her dedication to helping students transition into industry and graduate programs and for advocating for students in STEM. A Fellow of the Society of Women Engineers and a Senior Member of IEEE, among other recognitions, Anne continues to inspire through her teaching, research, and outreach. Her story is one of persistence, passion, and purpose, a testament to the power of early encouragement, hands-on experience, and the courage to forge a path where few women had gone before.

Foundations in Curiosity

Anne Lucietto's journey into engineering and later engineering technology began in a home where two rich cultural traditions German-American and Italian-American, intertwined to create a unique environment of discipline, creativity, and intellectual curiosity. Raised just outside Chicago, Anne was the eldest child in a family that valued hard work, education, and the joy of discovery.

Anne was naturally curious about how things worked from a young age. She was the child who could not resist taking apart toys, clocks, and household appliances, not to break them but to understand them. Anne would spend countless hours at the shop learning the basics of machining, observing the precision of metalwork, and absorbing the culture of craftsmanship. Rather than discouraging her mechanical interests, her father welcomed her into the shop, treating her curiosity as a strength, and teaching her how to use the machines and create metal parts.

Anne's explorations were not limited to the family business. She was a regular visitor to the local dump, where she would scavenge for discarded electronics, motors, and scrap metal. With a wagon tied to the back of her bicycle, she hauled her finds home to take them apart and study their inner workings. These adventures were more than childhood play, they were early engineering labs, where she learned through hands-on experimentation and problem-solving.

Her family also nurtured her love of learning in broader ways. Her aunt, Dr. Lena Lucietto, a dedicated educator, introduced Anne to the beauty of language and the power of communication. Whether it was practicing Italian phrases, exploring German roots, singing French songs, or diving into English literature, Anne's early exposure to multiple languages and cultures helped shape her into a thoughtful and articulate communicator. These skills would later serve her well in both academia and industry. In addition to intellectual pursuits,

Anne learned the value of handwork and craftsmanship from the women in her family. She spent cherished hours with her great-grandmother, grandmothers, mother (Elaine), and other friends and relatives learning traditional skills such as crocheting, sewing, quilting, knitting, tatting, and cooking. These experiences deepened her appreciation for creativity and patience and connected her to generations of knowledge and care passed down through loving hands.

Complementing these skills, Anne also learned technical skills such as part design and drafting, metalworking found in machine shops (including the family business), and woodworking from her father and other family members. These hands-on experiences fostered a deep respect for precision, problem-solving, and the satisfaction of building something tangible, skills that would later influence her approach to engineering and innovation. She continues many of these crafts today, including the one shown here, where she modified a pattern to suit a unique quilt.

Life brought its share of challenges. In third grade, Anne's family moved, and she had to adjust to a new school environment. The transition was tough—new classmates, unfamiliar routines, and the pressure of fitting in. But Anne adapted with quiet determination, channeling her energy into academics and extracurriculars.

She also excelled in athletics and the arts. In her school district, she consistently placed first or second in sprinting competitions, showcasing her drive and competitive spirit. Her talents also extended to music; she won a state-level vocal performance competition in her first year of high school and continued that achievement annually until she graduated, a remarkable achievement that highlighted her confidence and versatility.

These early experiences—tinkering, scavenging, performing, and adapting—laid the foundation for Anne's future in engineering. They taught her resilience, creativity, and the value of precision and expression. Most importantly, they showed her that being curious, bold, and different was not just acceptable but a strength.

A Change in Key

Anne began her college journey not in an engineering lab but on a stage and in a science lab—pursuing a double major in music and biology at the College of St. Teresa in Winona, Minnesota. Awarded a voice scholarship for her exceptional vocal talent, Anne initially envisioned a future in medicine. She planned to attend medical school, combining her love for science with a desire to help others.

At St. Teresa, Anne thrived in both disciplines. Her days were filled with biology labs and music rehearsals, balancing the precision of science with the emotional depth of performance. Yet, even as she immersed herself in these fields, a quiet question lingered: was this the right path?

That question came into sharp focus during a conversation with her father. Ever pragmatic and supportive, her father reminded Anne of a practical reality: her younger sister would be starting college soon, and the family could not afford to support two children in school indefinitely. "You need to be finished when your sister starts," he told her. It was a moment of clarity. Anne realized that the long road to medical school might not be feasible and reconsidered her options.

Engineering, a field she had grown up around and always been drawn to, suddenly made perfect sense. It offered the intellectual challenge she craved, the hands-on problem-solving she loved, and a more direct path into a stable and impactful career. With her father's encouragement and growing sense of purpose, Anne boldly decided to pivot.

She transferred to Marquette University and declared a major in mechanical engineering. It was a significant shift from the expressive world of music and the theoretical rigor of biology to the structured, applied engineering discipline. But for Anne, it felt like coming home. The machines, the materials, and the design challenges all spoke to the part of her that had once hauled scrap metal home in a wagon to take it apart.

This decision marked a turning point in Anne's life. It was not just a change in academic focus, it was a declaration of identity. She was no longer just the curious girl in the family shop or the singer with a scholarship. She was an engineer in the making.

Building a Life in Industry

After earning her mechanical engineering degree from Marquette University, Anne Lucietto was certain she was done with school. With the long hours, the pressure, and the exams, she had completed her degree and was eager to put her knowledge to work in the real world. Graduate school was the last thing on her mind.

She determinedly entered the industry, taking on roles that spanned mechanical, electrical, civil, chemical, and environmental engineering. Her work involved energy generation, nuclear plant construction, chemical processing, and materials fabrication. She quickly built a reputation for her technical expertise and ability to manage complex systems with precision and insight.

Anne met her future husband while working at a nuclear power plant. Their shared commitment to engineering and mutual respect for work laid the foundation for a strong partnership. Together, they built a life rooted in shared values and professional ambition.

Eight years into her career, Anne found herself at a crossroads. She had gained significant experience in engineering, but she began to see the value of understanding the industry's business side as she began to climb the corporate ladder. Despite her vow never to return to school, she enrolled in an MBA program. This time, her motivation was different, she wanted to lead, manage, and make strategic decisions that shaped the direction of her work. The MBA allowed her to expand her perspective and strengthen her ability to navigate technical and organizational challenges.

At this time, what began as a conversation with a city mayor turned into a 21-year journey in public service for Anne. Inspired by her husband's civil engineering and land development work, Anne developed a strong interest in land planning. When she expressed this to the mayor, she emphasized how her engineering background could benefit the city's planning commission. The next morning, while listening to the radio with her husband, she heard her appointment announced.

Anne served on the city plan commission for over two decades, most of that time as deputy chair and the final 4 years as chair. She collaborated with the city, county, and regional planning bodies to improve infrastructure and community relations. She led two major rewrites of the city's planning document and initiated a cross-community group that fostered trust and cooperation across municipal borders.

Her industrial experience proved invaluable in navigating complex discussions and leading with clarity. Some meetings were contentious, one even required police intervention, while others were collaborative and productive.

Around this time, Anne and her husband made the life-changing decision to adopt a daughter, an older child who brought joy, complexity, and a new sense of purpose to their lives. Balancing the demands of an engineering career with the responsibilities of motherhood was no small feat. Still, Anne approached it with the same determination and thoughtfulness that had guided her through every stage of her journey. Her daughter often accompanied her to work on days off, creating a little world under Anne's desk with a makeshift tent, books, and toys. These visits offered her daughter a rare window into the professional world, meeting colleagues, researchers, and industry leaders that most children would never have the opportunity to encounter. These moments strengthened their bond and gave her daughter a unique appreciation for the value of education, hard work, and the importance of community.

During this period, Anne also deepened her involvement with the Society of Women Engineers (SWE), which would become a cornerstone of her professional identity. What began as a way to connect with other women in the field quickly evolved into a meaningful leadership journey.

Anne first served as vice president and then president of the Chicago Regional Section, where she championed initiatives to support women in engineering, foster mentorship, and build a strong sense of community. Her dedication and strategic vision led to further leadership roles within the organization.

With an MBA in finance, Anne brought valuable expertise to SWE's financial operations. She joined the finance committee, eventually becoming its chair, and later served as treasurer on SWE's international board of directors. Her work helped strengthen the organization's financial foundation and expand its impact. Throughout this journey, her husband remained a steadfast supporter, recognizing the significance of these rare and meaningful opportunities.

Through her work in SWE, Anne found a platform to advocate for women in engineering and a network of peers who shared her values and drive. These experiences reinforced her belief in the importance of representation, mentorship, and creating spaces where all voices are heard. Anne extended this commitment beyond professional circles through extensive outreach efforts to inspire the next generation of engineers. She led and participated in programs reaching thousands of K-12 students, introducing girls and boys to the exciting possibilities of STEM careers. Her work included organizing hands-on activities, recognizing student achievements in math and science, and speaking at major events such as the statewide science fair, which drew nearly 10,000 attendees. Through these efforts, Anne became a visible and passionate advocate for STEM education, helping young minds see themselves as future innovators, problem-solvers, and leaders. She has been recognized for these continued efforts repeatedly.

Anne's years in industry were marked by personal and professional growth. She built systems, led teams, raised a family, and helped shape the future of engineering through her leadership and advocacy. While she did not yet know it, these experiences prepared her for the next chapter, which would bring her back to the classroom, this time as a full-time teacher, researcher, and mentor.

A New Chapter—Academia and Adversity

After nearly three decades in industry and part-time teaching at community colleges, Anne Lucietto began to feel a pull toward something different, something more lasting. Her experiences mentoring young engineers and teaching technical courses had sparked a passion for education. She found joy in helping students connect theory to practice and began to move toward a full-time career in academia.

Her time in the community college system had been enriching. She taught a wide range of courses, often to nontraditional students who brought their own life experiences into the classroom. Anne admired their grit and determination and took pride in helping them build confidence in their technical abilities. But as she looked ahead, she realized that to teach at a larger, research-focused institution that granted graduate degrees and shaped the future of engineering education, she would need a PhD.

Returning to school was not a decision she took lightly. She had once sworn off further education after completing her bachelors, and then again her MBA, but this time, it was different. This time, it was about purpose. She enrolled in a doctoral program in Engineering Education at Purdue University, determined to deepen her understanding of how students learn and how to better support those in applied engineering.

The first 2 years of her PhD journey were intense but invigorating. She immersed herself in research on student learning pathways, building on her industry experience and teaching background. But just as she began her third year, the most critical phase of her doctoral work, her world was shaken.

Her husband suffered a near-fatal stroke.

The news came suddenly, and the impact was immediate. Anne found herself navigating hospital visits, rehabilitation plans, and the emotional toll of uncertainty while trying to maintain her momentum in a demanding academic program. The balance she had so carefully constructed began to teeter. There were moments when she questioned whether she could continue, whether the dream of becoming a professor was worth the cost.

But Anne had never been one to back down from a challenge. Drawing on the same resilience that had carried her through industry, motherhood, and leadership roles, she pressed on. She adjusted her schedule, leaned hard on her support network, and found strength in her research, which reminded her of the importance of persistence, adaptability, and human connection.

Eventually, she completed her PhD, a milestone that marked academic achievement and personal triumph. With her doctorate, Anne transitioned into a full-time faculty role at Purdue University in the School of Engineering Technology. She found the perfect intersection of her passions: teaching, research, and mentorship. She brought with her the wisdom of industry, hands-on skills, the empathy of a caregiver, and the insight of a lifelong learner.

Anne's journey into academia was not linear or straightforward. It was shaped by unexpected turns, personal trials, and a deep commitment to making a difference. But in the end, it led her exactly where she was meant to be, in the classroom and helping the next generation of engineers find their paths.

Where We Are Today: A Legacy in Progress

Today, Dr. Anne Lucietto is not only a professor at Purdue University in engineering technology—she is also a mentor, a researcher, a community builder, and a changemaker in engineering education. Her phone is a hub of activity. It is full of texts, emails, and various calls from students seeking guidance, colleagues collaborating on research, and the quiet hum of ideas being transformed into action.

Anne's current work is rooted in a deep understanding of both the technical and human sides of engineering. She teaches courses that range from power generation and applied fluid mechanics to leadership and organizational change, blending rigorous technical content with real-world relevance. Her students do not just learn equations; they learn how to think critically, communicate effectively, and navigate the complexities of modern engineering workplaces. She is known for teaching painlessly, which is a feat.

Her research continues to focus on the educational experiences of engineering technology students. She leads projects such as those funded by the National Science Foundation and the Indiana Office of Energy Development, exploring topics such as the role of academic identity in student persistence, the impact of small modular reactors on energy infrastructure, and the issues students face in engineering and engineering technology programs. These projects are not just academic exercises but are designed to inform policy, reshape curricula, and create more interesting and engaging learning environments.

Anne's influence extends far beyond the classroom. She is a sought-after speaker at national and international conferences, a mentor to early-career faculty, and a leader in professional societies. Her work with the Society of Women Engineers (SWE) continues, and she remains a respected voice. She also contributes to ASEE, IEEE, and other technical organizations, ensuring that engineering education evolves alongside the industries it serves.

Despite her many accomplishments, Anne remains grounded in the values that shaped her journey: curiosity, resilience, and a commitment to service. She often reflects on her early days, hauling scrap metal in a wagon, taking apart televisions, and learning from her parents and family. Those experiences, she says, taught her not only how to build things but also how to build a life of purpose.

Today, she balances her professional responsibilities with a rich personal life. She continues to support her family, including her husband, whose recovery from a near-fatal stroke remains a powerful chapter in their shared story. She finds joy in mentoring students, many of whom remind her of herself, curious, determined, and sometimes unsure of where they fit in. She offers them knowledge, encouragement, empathy, and assurance that there is a place for them in engineering.

Anne Lucietto's story is far from over. She writes new chapters daily through her teaching, research, and unwavering belief in the power of education to transform lives. Her legacy is not just in the systems she has improved or published papers, the people she has inspired, and the doors she has helped open.

In a world that still struggles with learning STEM, Anne stands as a beacon of what is possible when talent meets tenacity and when one woman dares to challenge the status quo, not just for herself, but for everyone who follows.

Brittany Newell from Farm to Faculty

Brittany Newell

Early Inspirations and Educational Path

Brittany Newell grew up on a small farm in Southern Indiana. She loved the outdoors and always enjoyed math and science in school. However, her journey into the engineering world began with a spark of curiosity ignited during her middle school years. Encouraged by her parents, she participated in a minority STEM camp at the University of Louisville. This experience was pivotal, as it introduced her to the engineering as a profession. The camp provided hands-on activities and exposure to various fields within STEM, allowing Brittany to explore her interests and develop a deeper understanding of the possibilities within engineering.

The intricate mechanics of how things come together and worked captivated her, aligning perfectly with her agricultural background. Her upbringing on a farm involved practical problem-solving and a deep appreciation for how machinery and systems operated, which naturally led her to be intrigued by mechanical engineering.

Initially, Brittany was torn between her love for science, particularly biology, and her fascination with engineering. She contemplated a career as a pediatrician, driven by her desire to help others and her interest in biology. However, she ultimately decided to pursue biomedical engineering, a field that seamlessly blended her interests in engineering and medicine. Biomedical engineering offered the perfect combination of her passions, allowing her to work on developing medical devices and technologies that could improve patient care.

This decision led her to Purdue University, where she enrolled in the first fully accredited class in biomedical engineering. Purdue's reputation for excellence in

B. Newell (✉)
Purdue University, West Lafayette, IN, USA
e-mail: bnewell1@purdue.edu

A. Lucietto (ed.), *Innovating for Tomorrow: Engineering Technology*, Women in Engineering and Science, https://doi.org/10.1007/978-3-032-07645-8_9

engineering and its innovative programs made it an ideal choice for Brittany. The university provided a supportive environment, access to cutting-edge research facilities, and opportunities to collaborate with leading experts in the field.

While an undergraduate, she worked as an hourly undergraduate research assistant which introduced her to Purdue research facilities and infrastructure. She also worked as a mentor in the Purdue Twenty First Century Scholars program. The Twenty First Century Scholars program was an Indiana scholarship initiative for lower income students to pursue a college education at public Indiana colleges. She tutored her peer students and worked on initiatives to help combat the low retention rates of low income, often first generation students. During this time, it wasn't all education related initiatives, she also worked hourly jobs in the dining court and as a valet parking cars to help fund her education. These experiences taught her a strong work ethic and introduced her to research at Purdue and teaching. Her undergraduate research experiences had given her a taste of the challenges and rewards of scientific inquiry, and she was eager to delve deeper into her studies which led her to pursue graduate school. She started graduate school immediately post her undergraduate graduation.

She started graduate school in Purdue's Agricultural and Biological Engineering program. Her graduate studies were guided by Dr. Joseph Irjudayaraj, whose work on nanomaterials and cancer-based research deeply intrigued her. Professor Irudayaraj's innovative approach to using nanotechnology for cancer treatment resonated with Brittany's desire to make a meaningful impact in the medical field. This was what led Brittany to switch to the Department of Agricultural and Biological Engineering, where she could expand her knowledge and engage in interdisciplinary work. The department's focus on integrating engineering principles with biological systems provided a unique perspective that enriched her research.

During her master's program, Brittany Newell focused on groundbreaking research in targeted drug delivery for cancer treatment using functionalized nanoparticles. Her work combined advanced concepts from biotechnology and biological engineering, specifically aiming to create nanoparticles directly targeting cancer cells. (Book Newell et al., 2012; Panchapakesan et al., 2011; Wang et al., 2012) The primary goal was to deliver therapeutic agents precisely to affected tissues, minimizing the harmful side effects typically associated with traditional chemotherapy.

For her PhD, Brittany shifted her research focus while remaining in the same department, Agricultural and Biological Engineering but choosing a concentration in mechanical systems. She completed here PhD under the mentorship of Dr. Gary Krutz, diving into electroactive polymer materials. Her dissertation centered on developing polymer actuator systems, particularly o-ring seals that can expand autonomously in response to detected irregularities, addressing structural vulnerabilities(Krutz & Newell, 2020; Newell et al., 2013, 2017). This project was funded by Parker Hannifin, and required a multidisciplinary approach that blended materials science with mechanical engineering and explored the practical applications of electroactive polymers showcasing their versatility in a variety of applications. With industry support and a real-world timeline, Newell was able to complete her PhD degree in just 2 years.

Newell's PhD graduation

Overall, Brittany Newell's academic contributions during her master's and PhD were pivotal in advancing targeted drug delivery technologies and developing innovative electroactive polymer systems. These two research avenues represent a crucial integration of engineering, materials science, and medical applications aimed at enhancing therapeutic effectiveness while reducing adverse effects for patients.

Dr. Newell's work on electroactive polymers was particularly innovative, as it explored the use of smart materials that could change shape or properties in response to external stimuli. These materials had applications in various fields, including aerospace, automotive, and medical devices. Her research contributed to the development of more reliable and adaptable systems, enhancing the performance and safety of critical technologies.

Throughout her educational journey, Brittany's dedication to her studies and her ability to integrate diverse fields of knowledge set her apart. Her interdisciplinary approach allowed her to tackle complex problems from multiple angles, leading to innovative solutions that had a significant impact on her field. Brittany's passion for engineering and her commitment to continuous improvement through technological advancement continue to drive her work, making her a leading figure in the development of embedded sensors and actuators.

Newell spent two and a half years in industry post-graduation gaining real world experience and honing her skills before becoming a faculty member at the age of 28 at Purdue University in the School of Engineering Technology with a courtesy appointment in Agricultural and Biological Engineering. She started as a tenure track assistant professor in 2015. She achieved tenure in 2021 and is currently an Associate Professor.

Overcoming Challenges

Brittany's academic journey was not without its challenges. To help with the cost of her education, she had to balance her studies with work, holding a job throughout most of her college career. This experience, while demanding, instilled in her a strong work ethic and honed her time management skills. Brittany learned to manage her time efficiently, often juggling her academic responsibilities with her work schedule. Brittany also came from a rural farming town in Southern Indiana. She had 150 students in her high school graduating class. At Purdue she transitioned to a college with over 40000 students. She had to learn to navigate from being a big fish in a small pond to a small fish in an ocean. She found that it was critical to find a good group of friends at the university and clubs in which she could lead and find her niche. She joined the Biomedical Engineering Society where she had several leadership roles and led several events over the years. She was also a member of running club and the Society of Women Engineers.

As she moved to graduate school, academically, Brittany faced the typical hurdles of learning the ropes in research as she moved into graduate school. Transitioning from the structured classroom learning environment to the open-ended nature of research was a significant shift. She had to adapt to real-world situations where theoretical calculations did not always match experimental results. This process involved meticulous control of variables and a deep understanding of experimental design. Brittany's initial experiences in research were humble, as she started from washing dishes in a lab as an undergraduate research assistant, then transitioning to graduate school and eventually a faculty member. These early roles taught her the importance of hard work, precision, and attention to detail. These skills are valuable in life and research.

One of the most significant challenges Brittany faced throughout her career was the quintessential challenge of work-life balance. The demands of her work and schedule often required her to sacrifice time with family and friends. Brittany has found that the importance of prioritizing which she still views as a skill she is continuously trying to improve. Prioritizing includes your life and work goals and finding a balance in achieving both. Otherwise, it is difficult to be content with either aspect of your life. Navigating birth, child-care, kids sports and school activities while staying on top of her career is a significant hurdle. Newell and her spouse both maintain full time, high stress engineering careers, work on her husband's family farm, while raising four kids.

Brittany's ability to manage these challenges is rooted in her resilience, adaptability, and willingness to work hard and put in the effort. She developed strategies to manage her time effectively, prioritize her tasks, and maintain work-life balance. Her experiences taught her the importance of self-care and the need to seek support when necessary. Life truly requires a village.

Mentorship and Career Path

Mentorship played a crucial role in Brittany's career development. Dr. Gary Krutz, her PhD advisor, was instrumental in shaping her career trajectory. Despite her initial intention to enter industry after completing her PhD, Dr. Krutz continuously encouraged her to consider academia. His advice proved prophetic, as Brittany eventually found her way back to academia after a stint in industry. The demands of her industry job, coupled with the arrival of her first child, made her reconsider her career path. Dr. Krutz's timely suggestion to apply for an academic position led her to make her way back to Purdue as a faculty member. Brittany has had several very helpful mentors during her time at Purdue. They have helped her navigate university procedures, teaching, and understanding the academic research organization.

Brittany's mentors have played a crucial role in her career development, providing guidance, support, and encouragement at every stage of her journey. Their belief in her potential and their willingness to invest in her growth have been instrumental in her success. Brittany's experiences highlight the importance of mentorship in achieving academic and professional goals, and she continues to pay it forward by mentoring the next generation of engineers.

Current Role and Research Focus

Brittany Newell is an associate professor in the School of Engineering Technology at Purdue University, specializing in mechanical engineering. Her role is multifaceted, encompassing both research and teaching. She runs the Adaptive Additive Technologies Laboratory, where the focus is on the development of sensors and actuators focusing at the material level. The lab utilizes various 3D printing techniques, traditional fused deposition modeling (FDM), direct ink writing (DIW), and microprinting to produce embedded sensors and actuators. Brittany and her team create custom filaments and inks, balancing electrical and mechanical properties to achieve optimal embedded systems (Fan et al., 2022a, 2022b; Gonzalez et al., 2019, 2022; Hernandez et al., 2022; Moreno-Rueda et al., 2023; Simonson et al., n.d.; Vallejo-Melgarejo et al., 2019). They also integrate artificial intelligence and machine learning to develop dynamic feedback responses within their systems(Chamorro et al., 2022).

The Adaptive Additive Technologies Laboratory is a hub of innovation, where Brittany and her team work on cutting-edge projects that push the boundaries of material science and engineering. The lab's research focuses on creating smart materials that can respond to environmental changes, enhancing the functionality and reliability of sensors and actuators. By leveraging advanced 3D printing techniques, the team can design and fabricate complex structures with precise control over their properties.

One of the lab's key projects involves developing flexible sensors for healthcare applications(Moreno-Rueda et al., n.d.). These sensors can monitor physical therapy exercises for patients recovering from medical procedures or injuries, ensuring that exercises are performed correctly and frequently. The sensors provide real-time feedback to patients and healthcare providers, improving the effectiveness of rehabilitation and reducing the risk of complications.

Another significant project involves industrial health monitoring in collaboration with an industrial partner. Brittany's team uses predictive maintenance techniques to analyze data from production lines, leveraging machine learning and artificial intelligence to predict part quality and machine downtime. This project aims to enhance operational efficiency and reduce maintenance costs by identifying potential issues before they lead to equipment failure.

Brittany is also involved in an autonomous boat project with the Naval Surface Warfare Center and Trine University. This competition challenges universities to develop autonomous solutions for maritime maneuvering, aimed at combating drug trafficking. Her team's innovative approach earned them second place in the competition. The project involves equipping a 12-foot boat with a suite of sensors and developing algorithms for autonomous navigation and object detection. The goal is to create a system capable of identifying and intercepting drug trafficking vessels, enhancing maritime security.

Autonomous boat competition (**a**) Student and faculty team (**b**) Purdue's boat

Additionally, Brittany is working on bioastronautics work. These initiatives aim to enable long-term stays on the moon by developing methods for growing food and constructing habitat structures. The research involves exploring the use of 3D printing technologies to create sustainable living environments and support systems for astronauts.

On the teaching side, Brittany has taught several courses at Purdue, including Applied Thermodynamics and the freshman gateway course(Maynard et al., 2018; Moreno Rueda et al., 2023; Romero et al., 2024). Currently, she teaches the freshman gateway course, Gateway to Engineering Technology which is the largest enrollment course in the School of Engineering Technology. This course serves as an introduction to the different disciplines within engineering technology. Brittany's teaching philosophy emphasizes hands-on learning and real-world applications, preparing students to tackle complex engineering challenges. She integrates her research into her teaching, using her projects as case studies and examples to illustrate key concepts. This approach not only enhances student engagement but also provides them with valuable insights into the latest advancements in engineering technology.

Brittany's dual focus on research and teaching allows her to stay at the forefront of her field while inspiring the next generation of engineers. Her commitment to innovation and education is evident in her work, making her a respected and influential figure in the School of Engineering Technology at Purdue University.

Navigating Obstacles and Gender-Related Challenges

Brittany's transition from industry to academia was not without its hurdles. Coming directly from an industry position and having her PhD work funded by industry meant she was accustomed to industry standards and timelines, rather than the academic focus on grant writing and publications. This shift required her to understand and adapt to university systems, a process she navigated with the help of mentors such as Dr. Matthew Stephens, Dr. Richard Voyles, Nancy Denton, and Dr. Alejandro Magana. These mentors guided her through the grant writing process, helping her transition smoothly into her faculty role.

As a woman in male-dominated STEM field, Brittany has faced challenges. She often finds herself mistaken for other roles, such as a note taker, secretary or similar roles, rather than being recognized as the engineer in the room. This has required her to assert her position and expertise clearly. Brittany has developed strategies to address these misconceptions, such as confidently introducing herself and her role at the beginning of meetings and presentations. She ensures that her contributions are visible and that her expertise is recognized by her peers and colleagues.

Newell found that she noticed more challenges related to motherhood in comparison to being a minority, female in a male-dominated field (Morgan et al., 2021). Motherhood in general presents its own set of challenges while working in a professional environment. The demands of parenting, particularly during the early years of her children's lives, required Brittany to find a balance between her career and family responsibilities. She had to navigate the complexities of maternity leave, childcare arrangements, and the need for flexible work hours. Despite these obstacles, Brittany has managed to balance her career and family life, leveraging the flexibility offered by her academic position. Purdue University's supportive environment has been instrumental in this regard, providing accommodations for faculty members with family responsibilities.

Examples of these challenges span pregnancy through child-rearing. When she was pregnant in her tenure-track role as an assistant professor, a colleague told her she should hide her pregnancy as long as possible since she intended to go up for tenure. This notion was clearly outdated, as Purdue and most institutions offer a 1 year clock extension for tenure to parents. However, the research shows this time period is challenging for academic mothers more so than fathers. She has been asked numerous times if her intended pregnancies were accidental because she had a career. Challenges also occur while on maternity leave, courses are covered by the institution but it is more difficult to manage graduate students and research while on a leave as it is not directly assigned to another faculty member. Newell navigated these challenges by asking co-investigators to take over certain aspects of her work at that time. While choosing to nurse your child or not is a personal decision, this can also be a challenge to navigate. While in industry and academia it can be challenging to explain the need for nursing breaks to colleagues and find a clean and suitable place to do so Newell has also been unable to attend overseas conferences and events during pregnancy and the first year after child birth. She had to carve out time periods for nursing while in academia as well. Newell has also been unable to attend overseas events and longer domestic travel during this time period, because as a nursing mother you are quite literally the food supply for your baby. Newell has missed out on academic publications and events as a result of being unable to travel. These are challenges that can be difficult to explain or talk about as they are not common in a male-dominated field. Even after the first year, childcare in particular and illness is significantly challenging. It's key that employers offer flexibility for families with two working parents and reasonable expectations on working hours.

On the other hand, Brittany's experience as a parent has greatly influenced her teaching and approach with students. It offers her a different view of students and their families. It has taught her patience and compassion and this shows in her teaching. It also further highlights the need for work-life balance. She strives to be present for her children while managing her professional responsibilities, often working late into the night after her children have gone to sleep in order to meet her work demands. This flexibility is a crucial aspect of her role, allowing her to maintain a balance between her personal and professional life.

Brittany's journey highlights the importance of resilience, adaptability, and a strong support network in overcoming the challenges. Engineers are problem solvers and this spans all aspects of life. Engineers make it work through perseverance, adaptation, and hard work. Her ability to navigate these obstacles and her commitment to her career and family serve as an inspiration to others in the field. Brittany continues to mentor and support young engineers, particularly women, encouraging them to pursue their passions and overcome the barriers they may encounter.

Organizational Support for Work-Life Balance

Brittany's organization, Purdue University, is highly supportive of work-life balance for its employees. This support is evident in various ways, such as accommodating schedules for faculty members who need to drop off their children at school. Before having children, Brittany was comfortable teaching early morning courses, but this became challenging once she had to manage school drop-offs. Purdue's flexibility in scheduling has been beneficial, allowing her to balance her professional and personal responsibilities effectively.

The university's commitment to work-life balance is reflected in its policies and practices. Faculty members are encouraged to communicate their scheduling needs, and the administration works to accommodate these requests whenever possible. This flexibility has allowed Brittany to adjust her teaching schedule to better align with her family responsibilities.

Brittany's experiences at Purdue highlight the importance of institutional support in achieving work-life balance. The university's commitment to flexibility and understanding has enabled her to thrive professionally and personally. Brittany's journey serves as a testament to the positive impact of supportive work environments on the well-being and success of faculty members.

Staying Current and Essential Skills for Success

To stay current with the latest developments and achievements in her field, Brittany reviews a substantial number of academic papers and is frequently invited to review for conferences and journals. Her role as a guest editor for a journal keeps her updated with the latest research and trends in her area of expertise. This position involves evaluating submissions, providing feedback to authors, and ensuring the publication of high-quality research. Writing review articles and researching new industry projects also keeps her informed about emerging technologies and methodologies. These activities require Brittany to stay engaged with the scientific community, continuously expanding her knowledge and understanding of advancements in engineering technology.

Brittany emphasizes the importance of being a lifelong learner and solving problems, which is a fundamental output of engineering technology. She believes that the ability to adapt and learn continuously is crucial for success in the rapidly evolving field of engineering. Brittany's commitment to lifelong learning is evident in her proactive approach to staying informed and her dedication to professional development.

Brittany believes that essential skills for success in engineering technology include perseverance, a strong work ethic, and the ability to think outside the box. Perseverance is vital in overcoming challenges and setbacks, while a strong work ethic ensures consistent effort and dedication to one's goals. Brittany highlights the importance of interdisciplinary problem-solving, as real-world problems often involve multiple fields such as mechanical, electrical, thermal, and materials engineering. Effective communication and collaboration with experts in these areas are crucial for developing integrated solutions. Brittany's interdisciplinary approach allows her to tackle complex problems from multiple angles, leading to innovative and effective solutions.

Brittany also stresses the significance of creativity and innovation in engineering technology. Thinking outside the box enables engineers to develop novel solutions and push the boundaries of what is possible. Brittany encourages her students to embrace creativity and explore unconventional approaches to problem-solving. She believes that fostering an environment that encourages experimentation, and innovation is essential for advancing engineering technology.

In addition to technical skills, Brittany recognizes the importance of soft skills such as teamwork, leadership, and adaptability. Engineers must be able to work effectively in teams, lead projects, and adapt to changing circumstances. Brittany's experiences have taught her the value of these skills, and she integrates them into her teaching to prepare her students for the demands of the professional world.

Brittany's dedication to staying current with the latest developments and her emphasis on essential skills for success reflect her commitment to excellence in engineering technology. Her proactive approach to professional development and her focus on interdisciplinary problem-solving make her a respected and influential figure in her field.

Future Outlook for Engineering Technology

Brittany sees engineering technology becoming increasingly relevant in the future. The industry is undergoing substantial changes, with a greater shift to data-driven solutions, smart products, and Industry 4.0 initiatives. These transitions require new skill sets and higher expectations for the goods and services produced. The integration of advanced technologies such as artificial intelligence, machine learning, and the Internet of Things (IoT) is transforming the landscape of engineering, creating opportunities for innovation and efficiency.

Additionally, industry is facing a wave of retirements, leading to a loss of tribal knowledge and experienced leaders. This makes it even more important for new engineers to be independent thinkers who can fill these roles and meet the evolving needs of the industry. Brittany emphasizes the need for engineers to be adaptable and proactive in acquiring new skills, as the demands of the industry continue to evolve.

Engineering technology, with its emphasis on applied real-world experience, provides students with a solid foundation to build from. Brittany believes that this hands-on approach is crucial for preparing students to tackle complex engineering challenges. By engaging in practical projects and real-world applications, students develop the skills and confidence needed to succeed in their careers. Brittany's teaching philosophy aligns with this approach, as she integrates her research into her curriculum, providing students with valuable insights into the latest advancements in engineering technology.

In conclusion, Brittany Newell's journey in engineering technology is marked by her dedication to innovation, education, and professional development. Her multifaceted role as a researcher, educator, and mentor reflects her commitment to advancing the field and inspiring the next generation of engineers. Brittany's proactive approach to staying current with the latest developments, her emphasis on essential skills for success, and her vision for the future of engineering technology make her a respected and influential figure in her field. As the industry continues to evolve, Brittany's contributions will undoubtedly play a significant role in shaping the future of engineering technology.

References

B. Book Newell, Y. Wang, J. Irudayaraj, Multifunctional gold nanorod theragnostics probed by multi-photon imaging. Eur. J. Med. Chem. **48**, 330–337 (2012). https://doi.org/10.1016/j.ejmech.2011.12.036

J. Chamorro, L. Vallejo, C. Maynard, S. Guevara, J.A. Solorio, N. Soto, K.V. Singh, U. Bhate, G.V.V.R. Kumar, J. Garcia, Health monitoring of a conveyor belt system using machine vision and real-time sensor data. CIRP J. Manuf. Sci. Technol. **38**, 38–50 (2022)

J. Fan, N. Deneke, S. Xu, B. Newell, J. Garcia, C. Davis, W. Wu, R.M. Voyles, R.A. Nawrocki, Electric poling-assisted additive manufacturing technique for piezoelectric active poly (vinylidene fluoride) films: Towards fully three-dimensional printed functional materials. Addit. Manuf. **60**, 103248 (2022a)

J. Fan, B. Newell, J. Garcia, R.M. Voyles, R.A. Nawrocki, Effect of additive manufacturing on β-phase poly (Vinylidene fluoride)-based capacitive temperature sensors. Adv. Eng. Mater. **24**(11), 2200485 (2022b)

D. Gonzalez, J. Garcia, B. Newell, Electromechanical characterization of a 3D printed dielectric material for dielectric electroactive polymer actuators. Sensors Actuators A Phys. **297**, 111565 (2019)

D. Gonzalez, J. Garcia, R.M. Voyles, R.A. Nawrocki, B. Newell, Characterization of 3D printed pneumatic soft actuator. Sensors Actuators A Phys. **334**, 113337 (2022)

J.A. Hernandez, C. Maynard, D. Gonzalez, M. Viz, C. O'Brien, J. Garcia, B. Newell, T.N. Tallman, The development and characterization of carbon nanofiber/polylactic acid filament for additively manufactured piezoresistive sensors. Addit. Manuf. **58**, 102948 (2022)

G. W. Krutz, B. Newell, *Electroactive polymers, methods of manufacture, and structures formed thereof* (2020).

C. Maynard, B. Newell, A. Lucietto, W. Hutzel, J. Garcia-Bravo, *Applied Learning within Thermodynamics: A Perspective on Energy Concepts* (IEEE Frontiers in Education, 2018)

H.D. Moreno Rueda, K. Simonson, J. Richardson, E. Efendy, R. Webster, B. Newell, *Analysis of Student Motivation in an Introdcutory Engineering Technology Gateway Course* (American Society for Engineering Education (ASEE), 2023)

D. Moreno-Rueda, C. Maynard, J. Hernandez, T. Tallman, J. Garcia, B. Newell, 3D printed flexible gripper with capacitance sensing, in *ASME 2023 Conference on Smart Materials, Adaptive Structures and Intelligent Systems*, (2023), p. V001T04A007. https://doi.org/10.1115/SMASIS2023-110732

Moreno-Rueda, D., Narvaez, D., McCarthy, E., McCormack, L., & Newell, B. (n.d.). *3D Printed Flexible Tactile Sensor for Rehabilitation*. https://doi.org/10.1115/SMASIS2024-140240

A.C. Morgan, S.F. Way, M.J.D. Hoefer, D.B. Larremore, M. Galesic, A. Clauset, The unequal impact of parenthood in academia. Science Advances **7**(9), eabd1996 (2021). https://doi.org/10.1126/sciadv.abd1996

B.A. Newell, G. Krutz, K. Harmeyer, M. Holland, *Integrity Sensing with Smart Polymers and Rubber Components on Vehicles (i.e. Tires, Hoses, Seals)*, SAE Technical Paper 2013-01–0742 (SAE International, 2013). https://doi.org/10.4271/2013-01-0742

B. Newell, G. Krutz, F. Stewart, K. Pascal, Chemically pre-strained dielectric elastomers finite element analysis, in *Proceedings of SPIE - The International Society for Optical Engineering*, vol. 10163, (2017), pp. 129–137

B. Panchapakesan, B. Book-Newell, P. Sethu, M. Rao, J. Irudayaraj, Gold nanoprobes for theranostics. Nanomedicine **6**(10), 1787–1811 (2011). https://doi.org/10.2217/nnm.11.155

J.D. Romero, M.A. Feijoo-Garcia, G. Nanda, B. Newell, A.J. Magana, Evaluating the performance of topic modeling techniques with human validation to support qualitative analysis. Big Data Cogn. Comput. **8**(10), Article 10 (2024). https://doi.org/10.3390/bdcc8100132

Simonson, K., Valdes, A., Narvaez, D., & Newell, B. (n.d.). *Production and Validation of Non-Planar Electroplated 3D Printed Circuit Boards*. https://doi.org/10.1115/SMASIS2024-140173

L.D. Vallejo-Melgarejo, R.G. Reifenberger, B.A. Newell, C.A. Narváez-Tovar, J.M. Garcia-Bravo, Characterization of 3D-printed lenses and diffraction gratings made by DLP additive manufacturing. Rapid Prototyp. J. **25**(10), 1684–1694 (2019)

Y. Wang, B.B. Newell, J. Irudayaraj, Folic acid protected silver nanocarriers for targeted drug delivery. J. Biomed. Nanotechnol. **8**(5), 751–759 (2012). https://doi.org/10.1166/jbn.2012.1437

A Fay Barber-Dansby: The Spark of Inspiration

A. Fay Barber-Dansby

Inspired by Trailblazers

The African Proverb guides Fay,

> "Rain does not fall on one roof alone."

A. F. Barber-Dansby (✉)
Purdue University, West Lafayette, IN, USA

Purdue University Statewide, Anderson, IN, USA
e-mail: abarberd@purdue.edu

A. Lucietto (ed.), *Innovating for Tomorrow: Engineering Technology*, Women in Engineering and Science, https://doi.org/10.1007/978-3-032-07645-8_10

Madam C.J. Walker, Mary Jackson, and Dr. Gladys West have been those raindrops inspiring Fay throughout her journey in engineering technology. Gloria Webb-Williams highlighted these amazing women: Madam C.J. Walker developed and marketed a line of hair care and cosmetic products and made a fortune. Mary Jackson worked at Langley Research Center in Hampton, Virginia. Mary Jackson started her career as a computer at the segregated West Area Computing division in 1951. Mrs. Jackson, despite segregation, took advanced engineering classes and in 1958 became NASA's first Black female engineer. Dr. Gladys West, a mathematician, worked on the processing and analysis of satellite data that led to the development of GPS. Their work fueled Fay to persevere. She knew their challenges were far greater than hers (https://www.womensnetworkingalliance.com/post/black-women-who-have-changed-the-world).

Early Love for Mathematics

Fay's journey into the world of engineering began with a love for mathematics that was nurtured from an early age. Growing up in Anderson, Indiana, Fay, the oldest of three children (sister Robin 3 years younger and brother Thomas 8 years younger), grew up with very supportive college-educated parents. Her parents cultivated a household full of love, faith, culture, and service. Fay attended Highland High School, where she encountered an exceptional math teacher named Mr David Jack.

Mr. Jack's enthusiasm for the subject was infectious, and he made math both fun and exciting for his students. Fay thrived under his guidance, taking advanced courses like algebra, geometry, and advanced math, which were far beyond the basic requirements for graduation at the time. This early exposure to advanced mathematics laid a solid foundation for her future academic pursuits and instilled in her a deep appreciation for the subject.

A Chance Encounter

Despite her passion for math, Fay initially considered pursuing a career in law. However, a chance encounter with Mr. Forrest, the director of cooperative education at her high school changed the trajectory of her life. Walking down the hallway one day, Mr. Forrest stopped Fay, who recognized her talent in mathematics and asked had she considered engineering as a college major. Fay recalls, "In the hallway, I knew who he was. I did not know he knew me. For him to initiate that conversation with me in the 1970s, a white man in Indiana to care about my career and my future, still blows my mind." After doing her research, Fay knew that she would enjoy majoring in engineering because of math-centered courses and projects. Fay reflects, "And here I am over 45 years later still in an engineering technology space, because of that scroll down a high school hallway."

Discovering Engineering

Intrigued by the possibilities, Fay decided to explore engineering further. She received an invitation to attend a summer camp at Purdue University called the Minority Introduction to Engineering (MITE) program. Marian Blalock, Director of the Minority Engineering Program (MEP), led this camp, a respected figure on Purdue's campus, who is now an Ancestor. Fay fell in love with Purdue and the myriad options in engineering that Blalock presented. It was during this camp that Fay realized that industrial engineering was the perfect fit for her—combining her love for math with her people-oriented nature. The camp provided Fay with a comprehensive overview of the various engineering disciplines, allowing her to make an informed decision about her future career path.

Co-op Program at General Motors

Fay's journey continued to unfold as she received another opportunity, this time from General Motors' Delco Remy division in Anderson. They offered her a co-op program that she could participate in while attending Purdue. Fay started working at Delco Remy at the age of 16, gaining valuable experience in the engineering field. Although she noticed the lack of diversity in the engineering department, the support and encouragement from her colleagues were unwavering. They were proud of her achievements and eager to see her progress through her educational and professional journey. The co-op program allowed Fay to apply her academic knowledge in a real-world setting, providing her with practical experience that would prove invaluable in her future career.

Challenges at Purdue University

The transition from high school to college was not without its challenges. Fay's first year at Purdue University was a stark contrast to her high school experience. She entered Purdue with a 4.0 GPA but quickly realized that the academic rigor was far greater than she had anticipated. The professors at Purdue were renowned experts, even Nobel Laureates, and some of her textbooks were the very ones they had authored. Fay found herself struggling in her physics class; on one exam, the average score was a mere 30, and she scored a 13. This was a humbling experience for someone who had always excelled academically. The demanding coursework and expectations pushed Fay to her limits, forcing her to develop new study habits and strategies to succeed.

Support from NSBE and Mentors

Fay's initial doubts about her abilities led her to consider changing her major. However, her father encouraged her to persevere. The turning point came when Fay attended a meeting on campus of the National Society of Black Engineers (NSBE). The senior members of NSBE shared their own experiences of struggling during their first year and showed a trend line of their GPA improving over time. They reassured the first-year students that they were not alone and that it was possible to overcome these challenges. This support and encouragement were crucial for Fay, helping her to stay committed to her engineering major. The mentorship and camaraderie within NSBE provided Fay with a sense of belonging and a network of peers who understood her struggles. NSBE was founded on the campus of Purdue University in 1975.

Benefits of the Co-op Program

Fay also found solace in the co-op program, which provided a practical application of her studies and a break from the intense academic environment. The program allowed her to gain direct experience and see the real-world impact of her work. Fay recalls, "Surviving freshman year, and then moving straight into the co-op program, where I alternated between work and school each semester including summers. By the time I was sick of Purdue, it was time to go to work. By the time I got sick of work, it was time to go back to Purdue. So, for me, without the co-op program, I'm still not sure if I wouldn't have burned out." The alternating semesters between work and school helped Fay manage the academic stress and avoid burnout. The co-op program also provided Fay with a sense of purpose and direction, reinforcing her decision to pursue a career in engineering.

Graduate Studies at Columbia University

A few years after completing her undergraduate degree, Fay decided to pursue graduate studies. She attended Columbia University, a decision influenced by her desire to expand her career options. Fay recalls attending a meeting of The Consortium, which aimed to encourage people of color to pursue an MBA. Fay applied and received admission offers from Indiana University, University of Michigan, and University of Texas at Austin. Fay also applied to Columbia University with career considerations into investment banking. At that time, Columbia was not a member of The Consortium. Fay initially considered attending Michigan, but then Columbia came through with an admission offer. She was thrilled and decided to attend Columbia for the Ivy League experience and sample life on the east coast. The diverse student body at Columbia provided Fay with a rich and varied educational experience, broadening her perspectives and enhancing her understanding of global business practices.

Career at General Motors

Fay's career at General Motors restarted again after she graduated from Columbia University. She initially thought she would transition into high finance, but GM made her an offer she could not refuse. Fay recalls, "Job offers are coming in, and I'm on a leave of absence from General Motors and figured they would increase my salary by a decent amount. They made an incredibly competitive offer that I was not expecting." Fay's practical nature led her to consider the cost of living in New York versus Indiana. She decided to return to Indiana and continue her career with GM. She was now in a product marketing position which was still very technical, involving product management and technical analysis. Fay spent almost 20 years at GM in various capacities, including industrial engineering and product engineering. Her time at GM provided her with a wealth of experience and opportunities to develop her skills and expertise.

Highlights and Challenges at GM

One of the highlights of Fay's career at GM was her time in Albany, Georgia, where she worked on per diem. Fay recalls, "They were sending young engineers on per diem trips to work in remote facilities. I was assigned to our generator plant in Albany, Georgia in 1985. We were working overtime, and all my living expenses were covered, even getting my laundry done was paid for. I met some amazing people in Albany, Georgia and developed some lifelong relationships." Fay's career at GM was marked by great opportunities and experiences. She worked in various positions, including industrial engineering, product engineering, and marketing. However, as GM began to consolidate its divisions and components became commodities, the culture at Delco Remy changed. Fay recalls, "The community that I was used to in the Delco Remy culture just disintegrated. Consolidating divisions, felt like takeovers." The changing corporate culture and the shift toward commoditization presented new challenges for Fay, forcing her to adapt and find new ways to contribute to the organization.

At one point, GM offered Fay an opportunity to work in China, as a technical team member. However, the prospect of needing a nanny for her daughter made Fay reconsider her career path. She decided to switch to academia and began teaching at Ivy Tech Community College, focusing on math and student success courses. This transition allowed Fay to leverage her industry experience and passion for education to make a meaningful impact on her students.

Transition to Academia

Fay's transition to Purdue University began when she met Corey Sharp at Ivy Tech, who was taking on an opportunity to be the director of the statewide location for Purdue in Anderson. Fay recalls, "I was with Ivy Tech for a several years and he said, 'I'm going to take an opportunity to be the director of the statewide location for Purdue in Anderson, and I want you to come on my team with me.' I had been an instructor for high school algebra camps funded by a grant Corey received from the Ball Brothers Foundation to assist high school students with staying in good standing for their 21 Century Scholarship in Indiana. That is how we met—again, those thin threads. For me, it was coming back home. Teaching at a college or university was always on my bucket list. I figured would just stay within the community college structure. But to be able to come back home, so to speak, with Purdue, and then understanding the ecosystem of Purdue with the statewide location. I knew the Anderson location but was not aware of the other locations statewide. The Polytechnic system gives the main campus experience in multiple locations statewide, and for me, this is a perfect fit as an instructor."

Fay found the transition to Purdue Polytechnic to be a refreshing change. "There is a facility here in town. It is exciting that a person can earn a 4-year Purdue degree in their county or surrounding counties. Seeing the massive difference in what Purdue was and what Purdue is, and how the ecosystem is still growing, I am very excited, especially by Purdue in Indianapolis. Polytechnic values work experience, especially in the statewide community. Incorporating my industry knowledge within the coursework and sharing my many automotive examples are relatable with many students."

Fay's extensive industry experience allows her to bring a wealth of practical knowledge to her teaching. "We've got student led racing teams on campus here in Anderson, who compete in the Grand Prix on main campus. Looking at some of the build work they're doing and then engaging in an automotive discussion keeps me connected to the industry. I receive the respect of the students, knowing that I'm bringing all this applied knowledge to the classroom, and then understanding my goal is to assist them in being the best candidate in an interview process. I was a college engineering recruiter for a while, too, with General Motors."

Rewarding Project

When asked about rewarding projects, Fay reflected on both her industry and academic experiences. "I really enjoy working in the industry. When I first started, we were doing computer simulations of the flexible manufacturing systems in the building. We developed simulations to evaluate how the systems were running and modeling what we were seeing on the line in real time. We are looking at the bottlenecks and as a team with operator input, solutions emerged. I enjoyed that project."

Gender and Diversity in Engineering Technology

As a woman in engineering technology, Fay brings unique perspectives to the field. "I think women bring a different lens to a situation which broadens the lens for everyone. At the start of each semester, I conduct mini interviews with each student during our first class session to learn their life experiences including work history. This becomes a learning session for everyone in class. I state they are bringing a body of knowledge with them, and that they will have a voice in the discussions to see how much they can add to a particular concept we may be discussing. I have completed training with Purdue's Center for Instructional Excellence, (CIE). In some courses, I administer pre-testing of knowledge around the concept before I jump into the lesson."

Fay emphasizes the importance of representation and relatability. "I believe being able to be more encompassing and trying to maximize the input of everyone in the class besides myself. Brilliance resides in all communities, and everyone deserves an opportunity for that brilliance to be cultivated and shared with humanity. Our presence at high school recruiting events is necessary for believability for female students who may be saying, 'Should I consider this degree?' And then having someone there they can see that looks like them, gender-wise or ethnically, I believe it makes a difference."

Fay's commitment to community and human-centered design projects is evident. "I love doing community or human-centered design projects in Anderson. I grew up here in Anderson. It is very personal for me to see how students can get an intercultural experience. And then, appreciating the way we have transitioned from the School of Technology to Polytechnic. Intercultural experiences are an enriching component in a student's undergraduate experience. It can cultivate that heart of service as well as reminding students that they are gaining skill sets most people will not be able to acquire, and they can give themselves more purpose beyond what they may do in industry. I participated in the Purdue's inaugural Festival of Teaching and Learning and received the Blue-Sky Award for Excellence in Interdisciplinary and Inclusive Pedagogy."

Fay's journey from high school to her current role as a senior lecturer at Purdue Polytechnic in Anderson is a testament to her resilience, adaptability, and commitment to her goals. Her story highlights the importance of mentorship, support systems, and the ability to navigate challenges and make informed decisions. Fay's experiences in both industry and academia have equipped her with a unique perspective that she now shares with her students, inspiring the next generation of engineers and technologists.

Gender-Related Challenges

When asked about gender-related challenges in her career, Fay's response was candid and reflective. "How far back do we go with that? I think after a while we sit back and say, 'I got to pick my battles.' There's some you just let go there is some you don't. I remember I was working one Saturday at GM. This was early in my career, and I had a Purdue cap on. I get on the elevator. This guy jumped on the elevator, so he looks at me, says, 'So you went to Purdue?' I'm like, 'Yes, I did. And where did you go?' I cannot remember his response; he did not have anything else to say then. We got off the elevator and parted ways. But you know, you get that a lot."

Fay continued, "The same thing again. I was doing something one evening in the instructor's room at Ivy Tech, and there was another faculty member in there. My 17-year-old daughter was with me, I was helping her do senior stuff for college applications. This guy turns around, looks at me, 'So where'd you go to school?' My daughter turned around, looked at me and I'm sure starts thinking to herself, 'He does not want to hear this answer, and I know what my mom is going to do.' I am trying to evaluate myself. 'Okay, do this teachable moment with your daughter, you know, because you could curse the man out, or you could show, what Michelle Obama says? When they go low, go high.' I chose the high ground, and I said, 'Purdue engineering major, and you know that school in New York called Columbia for my graduate work.' He then turned red in the face, then I said, 'So where'd you go?' He said 'Ball State.' I said, 'Good school.' He then got up and left the room." Fay's daughter completed her undergraduate studies at Spelman College, majoring in Chemistry.

Fay's experiences with gender-related challenges were not limited to casual encounters. "The most frustrating would be, and I know we've all experienced this, I would be in meetings and the vast majority in the room were men. I am giving my input and suggestions for some problem-solving or brainstorming, and no one is acknowledging me. And then somebody else will say the same thing that they got from me. Then you hear 'That's a clever idea. Let us get that down. Let us put that down.' I am thinking to myself, 'How many more years am I going to deal with this?' When you're one of few, it is what it is, and they're still trying to grow."

Positive Changes in Leadership

Despite these challenges, Fay also witnessed positive changes in leadership during her time at Purdue Polytechnic. "But on the flip side, I was in a meeting on the main campus. At that time Dr. Andy Schaffer was leading our team statewide, and a professor there was leading the session, and I asked a question. I think he might have referred to me a gal or girl in his response to me. It was rude, but my goal was getting my question answered. I did not think much of it, because women deal with that response so often in majority male spaces. Andy was in that meeting. Andy comes for a visit to Anderson. It had to be a month or so later. He was conversing with someone in our hallway. I say hello, Andy then excuses himself for a minute from that conversation, and comes to me and says, 'I want to apologize for what happened in that meeting. I dealt with that faculty member that day, but I just wanted to give you a personal apology.' It took me a minute to even remember that meeting. I am now thinking to myself, 'This is a different group of leaders than I'm accustomed to.' And I feel the same way about Dr. Ken Burbank, our Department Chair,

School of Engineering Technology and Corey. They value everyone and have an inclusive mindset. It is refreshing and motivating. This is the best leadership, especially male leadership culture, I have ever worked in, and most of my career has been with male leadership. And I have experienced the total spectrum: Abrasive leadership in the auto industry to value-based leadership in Polytechnic."

Fay reflected on the contrast between her past experiences and the current leadership at Purdue Polytechnic. "In the auto industry, they led by fear, and that was with everybody. Does that type of leadership toughen you up? It sure does. I concluded this is just the way it was—that I am going to always have this adversarial situation with male leadership, and I will get through it because at least I am adding value my skill set, otherwise they would not have hired me. Then to witness an entire 180-degree difference in leadership culture in Polytechnic, starting with our former Dean of Polytechnic Dr. Gary Bertoline, current Dean, Dr. Daniel Castro-Lacouture, Dr. Ragu Athinarayanan (my first department leader), Andy, and Ken. I did not think I would ever experience such an empowering group of male leadership at any time in my career. The culture we have within comes from leadership committed to their own personal growth and choosing to be a better people than what they were the day before and committing to encourage, empower, and positively challenge our diverse group of faculty and staff in our School of Engineering Technology."

Developing Grit and Resilience

Fay's experiences in the auto industry and academia have helped her develop grit and resilience. "Yeah. But the old days were... I had days at GM where I would sit in my car in the parking lot and think to myself, 'Do I really want to go in here today?' You feel like you're on an island by yourself for, you know, almost 20 years of that. I tell people, 'Yeah, I have grit. My comfort zone is expansive.'".

Advice for Young Women in Engineering Technology

When asked what advice she would give to young women considering a career in engineering technology, Fay's response was enthusiastic and encouraging. "Do it. Just do it. I mean, we are all living in technology. I love life today. The question is asked, 'What's the best time of your life?' It is today. With the amount of technology, and amazing work women are doing, you can find community. Be okay with the possibility of not having straight A's in every course in engineering technology as a student. It goes back to this, and I am always going to preach this too—it goes back to that Dr. Seuss book, 'Oh, the Places You'll Go.' It is what the degree has done for me, and I do not know where it may take me next, It is just so refreshing and free. You just feel free because you can pursue so many industries, move in between industry and academia, or consulting. Keep a mindset of innovation; most women already have this gift, whether you're putting your outfit together, changing your hairstyle, taking a piece of clothing, and totally reconstructing it—that's the foundation of what we do in engineering technology. And the applied side fits us. It truly does for me, even though I have a Bachelor of Science degree in Engineering, the application side of work and the projects you can engaged in, even as a student, are amazing. The service you can give back to community in a technical aspect will be needed to solve community challenges. You may still be one of the few women in your department, but opportunities for you will abound. The degree is a great foundation for professional schools like business, law, or medicine. Several classmates of mine chose those paths."

Fay's journey from high school to her current role as a senior lecturer at Purdue Polytechnic in Anderson is a testament to her resilience, adaptability, and commitment to her goals. Her story highlights the importance of mentorship, support systems, and the ability to navigate challenges and make informed decisions. Fay's experiences in both industry and academia have equipped her with a unique perspective that she now shares with her students, inspiring the next generation of engineers and technologists.

The Path into Engineering Technology and Beyond

Beth McGinnis-Cavanaugh

This chapter traces the inspiring and unconventional journey of Beth McGinnis-Cavanaugh, from a young woman uncertain of her academic future to an award-winning engineering educator and community college advocate. Influenced early by her father's belief in the power of an engineering degree, Beth's path included dropping out of college, raising a family, and returning to school at Springfield Technical Community College (STCC) with renewed purpose. Her experience at STCC was transformative, leading her to pursue advanced degrees and eventually discover a passion for teaching.

Beth's professional career took shape unexpectedly when she was invited to teach a statics class at STCC during graduate school. That experience ignited a lifelong dedication to education, particularly within the community college system. Alongside her teaching, Beth contributed to major NSF-funded STEM education projects and became a vocal advocate for diversity, equity, inclusion, and belonging in engineering.

A central theme of her story is her struggle with impostor syndrome—a challenge she faced throughout her academic and professional life. Rather than letting it define her, Beth used it to fuel her empathy and mentorship, creating a supportive environment for students who, like her, may have doubted their place in STEM.

The chapter also highlights Beth's unwavering belief in the value of community colleges. She champions them as critical spaces for innovation, access, and transformation, especially for underrepresented students. Her story is a powerful reminder that success in engineering is not confined to traditional paths—and that with perseverance, mentorship, and purpose, anyone can build a meaningful career in STEM.

B. McGinnis-Cavanaugh (✉)
Springfield technicalechnical Community College, Springfield, MA, USA
e-mail: bmcginnis-cavanaugh@stcc.edu

A. Lucietto (ed.), *Innovating for Tomorrow: Engineering Technology*, Women in Engineering and Science, https://doi.org/10.1007/978-3-032-07645-8_11

Introduction

Beth McGinnis-Cavanaugh's journey into engineering technology is not just a story of professional achievement, but it is a deeply personal narrative of resilience, reinvention, and the enduring influence of family. Her path was shaped early on by the quiet but powerful encouragement of her father, who, though not an engineer himself, came from a lineage of technically minded men and believed fervently in the value of an engineering degree. He often told Beth and her sister that an engineering degree was more than a credential—it was a passport to independence, lifelong employability, and economic mobility. In an era when women were still largely discouraged or redirected from technical fields, his belief in their potential was both radical and empowering.

Beth's early academic experience was not linear. Like many young people, she entered college enthusiastically but found herself unprepared for college life on every level. After just one semester, she left school, entered the workforce, and soon married and started a family. Yet, the intellectual spark her father had nurtured never dimmed. Years later, after the birth of her second child, Beth experienced a moment of reckoning. She realized she was not living up to her potential and boldly decided to return to school with a renewed sense of purpose.

Her return to education began at Springfield Technical Community College (STCC), where, after taking several math classes and an architectural drafting class at night, she enrolled in the engineering transfer program. Here, Beth began to rediscover her academic strengths and rebuild her confidence. She later transferred to the University of Massachusetts Amherst, where she completed her bachelor's and master's degrees in civil (structural) engineering. Along the way, she encountered challenges that tested her resolve—none more persistent than the quiet, internal voice of self-doubt.

Beth's struggle with impostor syndrome became a defining part of her journey. Despite excelling academically, she often questioned her place in the field, especially as a woman in male-dominated classrooms and professional spaces. These feelings, though painful, became a source of empathy and insight—qualities that would later shape her approach to teaching and mentorship.

Today, Beth is a respected educator at STCC and a passionate advocate for the transformative power of community colleges. Her work extends beyond the classroom into national conversations about equity and belonging in STEM, the importance of accessible education, and the need to support underrepresented voices in engineering. Her story is a powerful reminder that the path to success is rarely straight—and that sometimes, the most meaningful journeys begin with a detour.

Educational Journey

Beth McGinnis-Cavanaugh's educational path was anything but conventional, yet it was deeply rooted in the values and vision instilled by her father. A man of numbers and logic, her father was an accountant by trade but came from a family of engineers. He believed that an engineering degree was not just a professional credential but a symbol of intellectual rigor, adaptability, and long-term security. He often told Beth and her sister that with an engineering background, they could do anything, go anywhere, and never be without options. This radical belief, especially for daughters in the 1970s, laid the foundation for Beth's future.

Encouraged by this vision, Beth enrolled in college in the late seventies—a bold move when women in engineering were still a rarity. However, the transition was not easy. Like many college students, the academic environment felt rigid and impersonal, and Beth struggled to find her footing. She just was not ready. After just one semester, she made the difficult decision to leave. It was a choice that reflected not a lack of ability but perhaps a lack of maturity coupled with what she perceived to be excessive expectations.

In the following years, Beth entered the workforce, married, and became a mother. Yet, the intellectual spark her father had nurtured never faded. After the birth of her second child, Beth experienced a profound realization: she was not living up to her potential. She felt her mind was being underutilized, and the desire to challenge herself intellectually became impossible to ignore.

Beth enrolled at Springfield Technical Community College (STCC) with determination and a renewed sense of purpose. It was a pivotal decision. At STCC, she found a learning environment that was both rigorous and supportive. She began taking courses part-time, balancing her studies with the demands of raising a young family. She thrived academically in small classes that engendered meaningful personal relationships with professors and in an environment where almost every student had a "story". At STCC, she rediscovered her confidence as a student and refocused on engineering after taking multiple math courses and an architectural drafting course. Not content to settle for the lack of technical expertise in residential and other small structures, her interest in bridges and tall buildings led her to civil engineering. She enrolled in the engineering transfer degree program at STCC.

Given her success at STCC, where she earned her associate's degree in engineering transfer and was awarded the Joseph J. Deliso, Sr. Scholarship for Excellence in Engineering, Beth transferred to the College of Engineering at the University of Massachusetts Amherst. There, she completed her bachelor's degree and earned a master's degree with a focus on structural engineering. Her academic journey was marked by long nights, moments of self-doubt, and the constant juggling of motherhood and coursework. Yet, through it all, she persevered—driven by the belief that she could, and should, do more.

Her educational journey is a powerful example of how nontraditional paths can lead to extraordinary outcomes. It also underscores the importance of accessible, supportive institutions like community colleges, which provide a critical bridge for students who do not follow a linear, traditional college trajectory. For Beth, education was not just a means to a career—it was a path to reclaiming her identity, fulfilling her potential, and ultimately, finding her calling.

Professional Career

Beth McGinnis-Cavanaugh's professional career in engineering took an unexpected and transformative turn during her time as a graduate student. While pursuing her master's degree at UMass Amherst, she received a call from one of her former professors at STCC. A faculty member was going on sabbatical and someone was needed to cover their statics course. Beth, who had planned to design bridges and tall buildings upon graduation, had not considered a teaching career. But the pull to return to STCC was strong, and statics had been her favorite course as an undergraduate. She accepted the adjunct position for one semester.

What began as a temporary teaching assignment quickly became a defining moment in her life. Standing before a classroom, Beth discovered something she had not expected: she loved it. The experience of breaking down complex engineering concepts, engaging with students, and witnessing their "aha" moments lit a spark in her that she had not found elsewhere.

That single class marked the beginning of a new chapter. Beth continued teaching at STCC, first as an adjunct and eventually as a full-time faculty member. Over the years, she has taught various engineering courses like statics, dynamics, strength of materials, and structural analysis, as well as introductory physics (mechanics). Like any new instructor, she had her weaknesses, but there was a connection—a resonance—in her classrooms. She found herself energized by the classroom environment and the opportunity to inspire students. Her teaching style is grounded in empathy, clarity, and a deep understanding of the challenges facing nontraditional students, many of whom, like her, balance academics with work and family responsibilities. One student wrote

> There is a sort of admirable dissatisfaction she brings to practice; she is always trying new classroom tactics and learning from both success and failure. In her courses, I have done everything from testing the material properties of candy to competing in a national competition. I can vividly picture how she confidently guided a class of students who hardly had a strong handle on algebra to designing code-compliant concrete mixes made of recycled materials to be used in a parking structure.

Beth's influence extends far beyond the classroom. From 2008 to 2022, she co-led three National Science Foundation (NSF) research projects in collaboration with educators from the Smith College Picker Engineering Program. These initiatives focused on developing multimedia engineering education curricula for middle and upper elementary students to introduce young learners—especially girls and underrepresented minorities—to engineering. These curricula applied cutting-edge educational approaches like Imaginative Education (IE) and Knowledge Building (KB). Through these projects, Beth helped secure nearly $7 million in funding and created innovative learning experiences that allowed children and pre-college teachers to experience engineering in new and exciting ways.

Her work on these grants also allowed her to collaborate with leaders in the learning sciences and their application in engineering education. Although she does not hold a PhD, Beth immersed herself in the world of academic research, learning about curriculum development, pedagogical theory, and the nuances of educational assessment. These experiences enriched her teaching and broadened her perspective on the use of story as a framework to develop deeper understanding of engineering concepts and the societal impact of engineering, as well as engage students typically outside of the engineering pipeline.

For her work in engineering education at all levels, Beth was named the 2014 Carnegie Foundation for the Advancement of Teaching and the Council for Advancement and Support of Education *Massachusetts Professor of the Year*. In 2015, she received the STCC Scibelli Endowed Chair for Faculty. In 2018, she was named Outstanding Faculty Member at STCC and won the President's Award for Diversity and Inclusion. Additionally, she was the 2018 Faculty Mentor for the Massachusetts DHE *29 Who Shine* Awardee and earned the 2020 Journalism Award from the Boston Society of Civil Engineers Section (BSCES) for her articles and efforts at raising awareness of engineering.

Her professional career is a powerful example of how unexpected opportunities can lead to profound personal and professional fulfillment. What began as a temporary teaching gig evolved into a lifelong vocation. Today, she is not only a respected, award-winning educator but also a mentor, curriculum developer, advocate for inclusive STEM education, and a champion for community college women in engineering. Her journey underscores the idea that engineering is not just about building structures—it is about building people, communities, and futures.

Struggle with Imposter Syndrome

> I know Beth is advocating for and encouraging other women just as she does for me. Her mentorship and support began when I was her student and continued even after I was no longer at STCC. … I also witnessed many interactions she had with my peers. Sometimes she would let them know she reached out to a certain professor for a work-study opportunity; other times, she had written referrals for employment. In all of these interactions, it was clear how much of an advocate she is for others. She has advised me during difficult times--when I felt that giving up on pursuing engineering would be the easiest thing to do--and always helped me to see the silver lining.

For much of her academic and professional life, Beth McGinnis-Cavanaugh wrestled with a persistent and deeply internalized sense of impostor syndrome. Despite her academic success—earning top grades, completing a rigorous civil engineering program, and later excelling in graduate school—Beth often felt like she did not truly belong in the spaces she occupied. It was as though her accomplishments were the result of luck or timing rather than her own intelligence and hard work.

This feeling was acute at the University of Massachusetts Amherst, where she transferred after earning her degree at STCC. As a nontraditional student, a mother, and often the only woman in the room, Beth was surrounded by younger, traditional male peers who seemed more confident, prepared, and accepted. These types of remarks fed a narrative she had carried for years—that she was not smart enough, that she was somehow faking it, and that eventually someone would find out.

Beth's impostor syndrome was not just a whisper in the back of her mind—it was a constant companion that shaped how she approached her studies, work, and even her self-worth. She recalls nights working on complex assignments until the early morning hours, often undone by the overwhelming fear that she was not cut out for this path. She juggled soil mechanics homework at hockey practice and statics problems at basketball games and still questioned whether she was doing enough or doing it right.

These feelings were compounded by the subtle and not-so-subtle messages she received from her environment. In graduate school, she encountered male classmates—some graduates of elite institutions like the United States Military Academy (West Point)—who openly questioned the presence of women in engineering. Their outdated views and dismissive attitudes only deepened her sense of isolation and self-doubt. More than once, she was asked, "What are you doing here?"

But Beth's story is not just one of struggle, but it is also one of resilience and transformation. Over time, she began to recognize that these powerful feelings were not unique to her. They were part of a broader pattern experienced by many women and underrepresented individuals in STEM. Beth began reframing her narrative with the support of mentors like Dr. Linda Meccouri, a compassionate and visionary educator at STCC and champion of underrepresented women in STEM. Linda helped Beth see her strengths, potential, and most importantly, her right to be exactly where she was.

Beth's experience with impostor syndrome has become a cornerstone of her teaching and mentorship. She speaks openly about it with her students, especially those who, like her, are navigating unfamiliar academic terrain. She encourages them to acknowledge their doubts but not be defined by them. In doing so, she creates a classroom culture where vulnerability is not a weakness but a step toward growth.

> Beth made … resources so readily available to us, even providing lunch so that we would not need to miss (or pay for) a meal. She planned extensively to make sure we were getting access to all the necessary information. No matter how busy she was--and she was usually incredibly busy—she made sure to hold these meetings in order to give each of us an opportunity to reflect on the challenges we were facing at the college. I remember bringing up in a meeting how homework problems were structured around stereotypical masculine hobbies and therefore male students had an advantage beyond what was being taught. She listened carefully and advised me to reach out to the professor on the matter. Her attention and concern gave me confidence and validation--and I did reach out to the professor.

Today, Beth continues challenging the myth of the "perfect" engineer. She reminds her students—and herself—that engineering is not about having all the answers but about asking the right questions, learning from failure, and showing up with persistence and purpose. Her journey through impostor syndrome is not over, but it has evolved into a source of empathy, strength, and connection.

Community College Is Where It Is At

For Beth McGinnis-Cavanaugh, community colleges are not just learning institutions, but they are engines of opportunity, equity, and transformation. As a former student and long-serving faculty member at Springfield Technical Community College (STCC), Beth has firsthand experience with the profound impact these institutions can have on individuals and communities. Her advocacy for community colleges is not theoretical; it is rooted in lived experience, professional dedication, and a deep belief in the mission of accessible education.

Community colleges are harbors of vertical mobility, and broadening access and opportunities for students is something Beth considers a sacred charge. When Beth first approached student Menachem Sahler about the prestigious Jack Kent Cooke (JKC) Undergraduate Transfer Scholarship, he was downright cynical. "Those places [colleges] aren't for people like me", he said. Menachem—an intelligent, hardworking, yet jaded high-school dropout who was giving education "one last fling" at STCC—embodied the community college mission. His selection as a JKC Transfer Scholar—STCC's first—was a highlight of her career. "Congratulations on another successful student", wrote then STCC President Ira Rubenzahl to Beth. Menachem himself wrote:

> "I dare not ask how many hours Professor McGinnis-Cavanaugh selflessly invested in my academic success. I have little doubt that being awarded the Jack Kent Cooke Transfer Scholarship can be attributed as much to her tenacity as to my own academics. Without the JKC scholarship, I would likely not have been able to secure enough loans to attend RPI. With it, I was able to avoid loans entirely, loans that distract and demoralize too many of my peers. The lack of financial stress enabled me to focus on academics in a very demanding setting. Professor McGinnis-Cavanaugh introduced me to the JKC Transfer Scholarship and helped me navigate an absurdly involved application process. She represented me in front of a somewhat skeptical STCC nomination committee. At one point, she even set up a meeting with the school president on my behalf and recruited other teachers to help as well. I am focused and hard-working, but self-confidence is not my strength. However, Professor McGinnis-Cavanaugh believed in my ability to succeed and tirelessly encouraged me to try and try again. Being a truly great classroom educator is well and good but a dedicated investor in students' long-term success is something unique and to be highly valued."

Beth often speaks about the stigma surrounding community colleges—a perception that they are somehow "less than" 4-year universities. She challenges this notion at every turn. In her view, community colleges are where some of the most important and innovative educational work is happening. They serve students from all walks of life, including first-generation college students, working parents, veterans, immigrants, and those who have struggled in traditional academic settings. These students bring with them a wealth of life experience, resilience, and determination that Beth believes are essential to the future of engineering and technology.

Her journey is a testament to the power of the community college model. At STCC, Beth rediscovered her academic confidence, found mentors who believed in her, and laid the foundation for her future success. Now, as a professor, she works tirelessly to create that same environment for her students. She understands their pressures—juggling jobs, family responsibilities, and coursework—and meets them with empathy, flexibility, and high expectations.

Beth's commitment to community colleges extends beyond the classroom. She has been actively involved in national organizations such as the Society of Women Engineers (SWE) and the American Society for Engineering Education (ASEE), where she advocates for greater recognition of the role community colleges play in STEM education. She co-leads the SWE Community Colleges Affinity Group, chairs SWE's Academic Working Group, and sits on the SWE Magazine Editorial Board—all to amplify the voices and experiences of community college faculty and students.

Beth is deeply passionate about the vital role community colleges play in advancing diversity, equity, and inclusion in STEM. She often quotes William Wulf, former president of the National Academy of Engineering, who championed the value of both collective diversity—race, ethnicity, and gender—and individual diversity–life experience and background—as essential to "good engineering". This dual dimension of diversity, she notes, is embodied in the community college student population. She argues that community colleges are uniquely positioned to prepare students to bring diverse perspectives to engineering, resulting in stronger technical solutions and more meaningful societal impact.

Beth's vision is clear: community colleges are not a fallback or a second choice—they are a vital, vibrant part of the educational ecosystem. They are where innovation meets access, where potential is nurtured, and where the next generation of engineers, technologists, and leaders is being built from the ground up.

Conclusion

Beth McGinnis-Cavanaugh's journey is a powerful example of how perseverance, mentorship, and a passion for education can lead to success in the field of engineering technology. Her journey fuels her advocacy for community college students as critical for STEM equity and diversity. She regularly presents and publishes about community college STEM pathways; nontraditional, nonlinear STEM pathways; first-generation experiences in STEM; and bias against community colleges as a class issue.

Beth's story highlights the importance of community colleges, the impact of overcoming imposter syndrome, and the value of advocating for women in STEM. Beth continues to inspire and support the next generation of engineers through her teaching, mentorship, and advocacy. Creating communities that value and leverage the intersectional experiences of community college women in engineering is a cornerstone of her work.

> Beth has been one of the most important mentors in my life and has without question shaped my professional future. I know the same is true for many of her other mentees. She has always exemplified all four dimensions of mentoring, and her advocacy work has and will continue to make a difference for community college women in STEM. I, for one, am hugely grateful for her guidance and care. My admiration and appreciation for Beth's efforts for all of us is immense.

Breaking Barriers: A Woman's Journey Through STEM and Cultural Resistance

Jérika Lamas

My name is Jérika Lamas, and I am a Peruvian born in Brazil. I grew up far from my extended family in a warm but strict household. From an early age, I witnessed my parents' dedication and resilience as they worked tirelessly—not only to be present in our lives but also to provide us with the best opportunities for the future. Their hard work and sacrifices shaped my understanding of perseverance and ambition.

I attended a private school until the age of 13. In Brazil, private education is expensive, but it also provides excellent opportunities for the future. I was a good student, particularly in science, though I showed little affinity for languages. My father understood and supported my strengths, encouraging me to focus on what I enjoyed most.

At 14, my parents decided that a technical school would be my best path. This choice would give me the opportunity to enter the workforce immediately after graduation. The curriculum was rigorous, with classes in the morning, afternoon, and sometimes evenings or on Saturdays. It was challenging, but those years were fantastic—I earned a degree in electronics and made lifelong friends. At the time, I had no real awareness of what technical education meant for girls; I simply found it all fun and exciting.

During that period, my mother, an English teacher, insisted that it was time for me to learn English. My father initially opposed the idea—he did not see the need for it in my future career. But my mother was persistent and refused to let him have the final say. Looking back, I am grateful for her determination. I now believe my father's resistance was not just about practicality; perhaps he feared that learning English would open doors for me to leave Brazil, as he had done himself, leaving his own country for a better future. And that would mean having his children far from home.

J. Lamas (✉)
Purdue University, West Lafayette, IN, USA
e-mail: jerika.lamas@sirris.be

A. Lucietto (ed.), *Innovating for Tomorrow: Engineering Technology*, Women in Engineering and Science, https://doi.org/10.1007/978-3-032-07645-8_12

At 17, I started university. Since both my father and my oldest sister were engineers, it felt like a logical path to follow. My father was a chemical engineer, and my sister pursued biochemistry. I, however, preferred physics over chemistry, which led me to choose Materials Science and Engineering.

University life was not as enjoyable as technical school. There was a strong sense of competition, likely because graduation marked the transition into the workforce. For the first time, I was confronted with a new reality—I was one of the few women in a class dominated by men. In this competitive environment, making friends was not easy. Perhaps it was because I ranked among the top three students in my class, which was sometimes met with envy. I worked incredibly hard to maintain my high grades, and my efforts paid off—I earned a scholarship to conduct scientific research in my spare time. This was a fantastic opportunity, as it not only provided financial support but also allowed me to learn and grow academically.

Through my research work, I connected with senior students and built friendships beyond my own class. I loved the research I was doing, and that experience solidified my decision to stay in academia. I was convinced that teaching and conducting research in my field would make me happy. I had built a strong network and developed close relationships with my professors, who taught me invaluable lessons.

I am especially grateful to my two supervisors. They not only guided me in superconductivity and programming but also taught me how to write impactful research papers and present effectively to large audiences. More importantly, they showed me that kindness can take you far in your career.

In my final year of university, I had already collected a significant number of scientific papers in superconductivity and presented my research at several national conferences. My plan was clear: after completing my master's degree, I intended to pursue a PhD in the United States. The future seemed bright.

At the end of my studies, I sought an internship abroad and spent 2 months at a research institute in Slovenia, working on magnetic caloric materials. This experience opened my eyes to how far one could go in science. Even more inspiring was the fact that I was surrounded by strong, determined women who had fought hard to secure their place in the field. It was an environment unlike any I had experienced before.

During this period, I met someone who would change my path. When I returned to Brazil to complete my degree, I continued with my master's for another year and a half, in addition to the 5 years of university. But instead of heading to the United States as originally planned, I decided to pursue a PhD in Belgium to be with my boyfriend. He had graduated as a mechanical engineer from Ghent University and introduced me to research groups that aligned with my interests. I had two options: one in Materials Engineering and another in the Physics Department. I chose physics because it focused more on experimental work.

In 2009, I moved to Belgium. I gathered all the money I had, and with a heavy heart, I said goodbye to my family to embark on an unknown journey abroad.

Starting over in a completely different country is not as easy or exciting as it might seem. It feels like being trapped in a surreal dream where everything is unfamiliar, and you cannot find your way out. Everything was different—the people, the culture, the food, the weather, and the social norms. The cultural differences were more profound than I had anticipated. And to put it bluntly: I hated it here.

Unfortunately, I became a target of discrimination and racism. The stereotypical jokes, disguised as harmless humor, were deeply hurtful. At first, wanting to be polite and fit in, I did not react. I suffered in silence. I made every effort to integrate—I learned Dutch, adapted to local customs, and tried to blend in. But it was made clear to me that I would never truly belong. Whenever I expressed discomfort, the response was often the same: "Just go back to your country."

Over time, I learned to stand up for myself. If a comment was offensive, I made sure to call it out, refusing to let it harm me anymore. Even now, 15 years later, I still face prejudice. But today, I react. I no longer stay silent.

In my early years in Belgium, my only goal was to finish my PhD and leave—to escape back home. Despite all the challenges, though, Belgium taught me a great deal about myself and deepened my love for my own country. I began to appreciate even the smallest things I had once taken for granted. Over time, I also learned to accept cultural differences—to a certain extent. When people reacted poorly to cultural misunderstandings, I took it upon myself to explain. After all, if we, as foreigners, do not help them understand, how else will they learn?

My PhD in Solid State Science was both mentally and emotionally draining. My research focused on using magnetic sputtering to deposit yttria-stabilized zirconia as an electrolyte for fuel cells. The first 3 years were intense but rewarding—I conducted as many experiments as I could, met some great people, published my research, and grew both personally and professionally. I even received a recognition of my work by the European Materials Research Society E-MRS in 2011 and the Society of Vacuum Coaters in 2013.

But my final year was beyond challenging. I had to stop the experimental work I loved to develop a theoretical model—completely outside my comfort zone. On top of that, writing my thesis became a nightmare. No matter how hard I worked, nothing I wrote seemed to satisfy my supervisor. I felt worthless, as if everything I had done was meaningless. I was mentally exhausted.

By the time I completed my PhD, I knew I did not want to stay in academia. It was painfully clear that only a select few had the chance to become assistant professors, and as a foreigner, I had no real shot.

One moment that has stuck with me is something a colleague once said to me. He asked if I had a boyfriend. When I answered yes, he replied, *"Lucky you. Men don't like smart women."* I was speechless. Was I supposed to feel flattered for being considered smart? Or concerned about the way men perceive intelligent women?

During my PhD, I got married. My husband had career plans that required us to stay in Belgium a little longer, so he kindly asked me to remain for a few more years. Hoping to eventually return to my home country and become a professor, I took a postdoctoral position in the organic chemistry group at the Université Libre de Bruxelles, working on plasma coating—a chemical vapor deposition technique.

However, this new chapter brought its own set of challenges. Chemistry had never been my strongest subject, and on top of that, I now faced a language barrier. Until then, I had been working in English within a Dutch-speaking environment, but in Brussels, French was the dominant language. Fortunately, I could still work in English, but following everything else in French was difficult. It was impressive to see how many languages people spoke in such a small country.

Despite the language struggles, I found French-speaking colleagues to be incredibly warm and welcoming. I enjoyed my time in Brussels and Charleroi, and I wanted to continue working with them. After a year and a half, I received a job offer from a French-speaking company that collaborated with the university. Unfortunately, the position required me to commute 200 km a day, which did not seem feasible at the time.

Seeing that my husband was thriving in his career, I had to let go of my academic dream for good and transition into the industry. I moved back to the Dutch-speaking part of Belgium, settling in a small town, where I joined Bekaert as a product engineer in the metal industry. My role involved the technical management of small-diameter wires used in automotive applications, such as window regulator cords and heating cords for AdBlue systems.

The cultural shift was significant. The people were reserved but kind, reflecting the rural character of the region. However, once again, I found myself in a male-dominated environment—this time as the only foreign woman in the team. Although I spoke Dutch, I was never assigned Dutch-speaking tasks. Instead, my focus was on managing a production line in China and maintaining contact with European customers. This role had me traveling frequently, which allowed me to experience different cultures and broaden my perspective.

One of the highlights of this job was working under a great manager. He taught me how to navigate different personalities, adapt to various situations, and collaborate effectively with diverse teams. We had a strong professional connection, and he was not afraid to offer direct feedback for the sake of my personal growth.

I have always been straightforward and direct—qualities that are not always well received in a corporate setting. My manager once told me something I will never forget: "People may ask for honesty, but they don't really want to hear the truth." That statement stayed with me. I still believe that directness is often more accepted from men than from women, but this experience taught me the importance of choosing my battles. If a discussion is unlikely to lead anywhere, then it is not worth investing my time in it.

During my time at Bekaert, I became a mother for the first time. It was not something I had planned, so the experience was entirely new to me—and, to be honest, it hit me hard. At first, I believed I could continue working full-time, focusing on my career while simply fitting motherhood in between. But I quickly realized that I wanted to give my best both at work and at home, and no matter how hard I tried, I felt like I was failing at both.

Over time, I started experiencing boreout—the opposite of burnout, where I felt unchallenged, unmotivated, and disconnected from my work. I felt useless, as if my contributions had no real impact on the company. At the same time, I was missing out on precious moments with my daughter. She was the first to be dropped off at daycare at 7:30 a.m. and the last to be picked up at 6 p.m. Our time together was reduced to a routine: feeding her, giving her a bath, and putting her to bed. There was no real bonding, no quality time.

To make things worse, my frequent work travels meant I missed many of her "firsts." And as if that emotional weight was not enough, I also lost my cat to cancer—a pet I had considered like a child. The grief, the exhaustion, and the sense of detachment from both my job and my family became overwhelming.

That is when I made a decision: I quit my job. I needed something new, something that would excite me again. More importantly, I needed a job closer to home, with less travel, so I could be present for my daughter. With that change, I finally had the time to rebuild my connection with her, to recover from my boreout, and to process the loss of my beloved cat.

I started working at Morrow Optics, a startup developing multifocal lenses using liquid crystal technology. The concept behind these glasses was brilliant—an incredibly smart application of science to something truly useful. I was responsible for reliability testing and quickly advanced to managing the coating application by spray coating and inkjet printing. I loved being able to fine-tune processes and implement new technologies. I spent a lot of time in the clean room, working alongside amazing international colleagues, many of whom became lifelong friends.

Then COVID-19 hit. Like so many others, I was suddenly isolated—from colleagues, from friends, and most painfully, from my parents in Brazil. Travel restrictions in Europe were strict, and I went 1.5 years without seeing my parents and 2.5 years without seeing my sister. But despite everything, I am grateful for one thing: the time I got to spend working from home with the best colleague I could have—my daughter. We bonded in a way I never expected, and those months together were a gift.

However, when restrictions eased and we returned to the office, I noticed a drastic change in the company culture. The team had split into two distinct groups: R&D and Operations. I was in Operations, where we were overloaded with work, pressured by management with unrealistic deadlines, and held accountable for everything. Meanwhile, R&D moved at a slow pace, took no responsibility, and pushed their unfinished work onto us. I could not believe management tolerated this imbalance.

What shocked me even more was the toxic culture that emerged. Bullying, public shaming, and mistreatment became normal. The business was failing, and money was being spent recklessly. Over time, I discovered that this chaos stemmed from an affair between the CEO and an employee who manipulated him into prioritizing her decisions over everything else. Anyone she did not like was fired.

During this period, I became pregnant with my second child, and the stress at work made me physically ill. I did not want to return after maternity leave, but my family reassured me it would be okay. They were wrong. Things only got worse. I frequently clashed with the "chosen" group because I could not tolerate seeing people being belittled in front of me. Along with other employees, we reported the situation to the board of directors, but they did not care. We sought help from an external workplace safety institution, which acknowledged the toxicity but only advised us to find another job. Even a government agency that supposedly dealt with workplace bullying threatened me: "If we investigate and don't find enough proof, we'll sue you for false testimony."

I had proof—emails, slack messages—but the fear of not being believed was paralyzing. Some of my colleagues had work permits tied to their jobs, so they could not risk speaking out. It was a helpless situation.

Six months after my return, I quit. Many others followed. Out of 22 people in my team, only three remain at the company today. Eventually, the CEO and his affair were fired, and peace was restored—but the damage had already been done.

To this day, a small group of us—four ex-employees and one still at Morrow—meet once a month. What started as a kind of therapy group has become a close-knit circle of friends. We all went through trauma there, but I am incredibly proud of what each of them has achieved since. Slowly, we are all moving forward, leaving the negativity behind.

I was fortunate to find a job at Sirris, a research company located just 9 km from my home. The flexible working hours allow me to maintain a healthy work–life balance, and I can work from home whenever needed—whether it is a school-free day or one of the kids is sick.

My work at Sirris is divided into two key areas. One part involves applied corrosion research for industry, helping companies find solutions to their corrosion problems. The other focuses on corrosion monitoring for renewable energy, particularly offshore wind turbines, to extend their lifespan.

However, even in this role, I am once again confronted with sexism. I work in the metals group, and I noticed early on that it took time for my colleagues to fully acknowledge my expertise. I have to work harder to prove my worth—something my male colleagues do not have to do. At technical events, I still see how men have an advantage over women. Our opinions are often dismissed unless they are backed up by a senior person or a man.

From the projects I have worked on, I have observed that the ratio of women to men in this field is about 1:15. Sometimes, this reality frustrates me. I wonder: Are women genuinely not interested in this area? Or do they leave because they are tired of constantly having to prove themselves? I wish I could change things, but I often feel powerless against the deep-rooted biases in this industry.

I truly believe that if women continue to strive for equality, there will no longer be a distinction between "male" or "female" jobs. Technology, whether in the form of machinery or AI, has the potential to level the playing field for both genders.

Facing discrimination, a toxic environment, or sexism should only make us stronger. These challenges are like scars left from falling—they hurt, but they should never stop us from reaching our goals. We must never be afraid to ask for help, whether it is through therapy, the support of friends, or simply taking time for ourselves.

And if you see another woman struggling, do not hesitate to offer your support. Showing them they are not alone helps lighten their burden and makes the pain a little more bearable.

While my story may not seem entirely positive, I am incredibly proud of myself for having made it through everything with a reasonable level of mental health intact. I have a wonderful family, a great group of friends, and ex-colleagues who trust me and are always there when I need them. I now work 90%—a choice that gives me more quality time with my daughters.

I have always been involved in several social projects. I teach summer lessons to children who are either mentally disabled or come from socio-economically disadvantaged backgrounds. I show them the applications of physics through experiments and games. This work motivates me even more to help them build a better future. At the same time, these children give me strength to continue doing what I do best. They teach me things I was fortunate not to experience, making me more generous and grateful for what I have.

There is no greater healing than turning our own struggles into a source of strength for others. By supporting those who face the same challenges, we create a ripple effect of compassion and empowerment, helping to build a future where no one has to endure what we once did.

Engineering the Unseen: Trailblazing Path in Robotics

Upindar Kaur

From a young age, Upinder Kaur harbored an extraordinary dream—to become an astronaut. While most children her age were immersed in cartoons and playground games, she was captivated by the stars. The quiet corners of her school library became a gateway to the cosmos, filled with books about space travel, the history of NASA, and biographies of astronauts who had defied gravity and expectations alike. These stories didn't just fascinate her, they sparked a deep sense of purpose.

Growing up in the bustling heart of Delhi, India, a career in space exploration might have seemed ambitious. While the infrastructure for aeronautics was still developing, and such ambitions were uncommon, especially among young girls, Upinder's fascination remained strong. As she explored into the journeys of those who had reached the stars, she noticed that many had started as mechanical engineers.

This realization became her north star. To her, mechanical engineering was not just a discipline, it was the launchpad to her dreams. It combined logic, creativity, and the potential to build real-world systems, whether machines on Earth or rovers bound for Mars. She understood the road would be challenging, particularly in a field where she would often be the only woman, but she felt compelled to follow this path.

U. Kaur (✉)
Purdue University, West Lafayette, IN, USA
e-mail: kauru@purdue.edu

A. Lucietto (ed.), *Innovating for Tomorrow: Engineering Technology*, Women in Engineering and Science, https://doi.org/10.1007/978-3-032-07645-8_13

Building the Foundation

Her childhood was a canvas for imaginative engineering experiments, long before she stepped foot in a formal lab. While other children played with dolls or board games, Upinder's world was built from paper, glue, and curiosity. She meticulously assembled intricate paper models of NASA's space shuttles, each fold and tab carefully aligned, transforming simple sheets into marvels of aerospace design. These were not just toys to her—they were studies in form and function, representations of distant planets and orbital mechanics she yearned to understand.

At home, her younger brother received wooden kits meant to inspire creativity, including motorcycles made from laser-cut shapes, waiting to be pieced together. But more often than not, those kits found their way into Upinder's hands. Driven by creativity, she found each kit a source of joy and ingenuity. She worked hard on each such toy, meticulously building it part by part.

In school, she joined the Astronomy Club, where she began to explore the universe through a more scientific lens. There, she learned to assemble telescopes and build working sundials from paper and cardboard, teaching herself how to read shadows and measure time by the sun's arc. These tools, though simple, tapped into the ancient roots of scientific discovery, and she reveled in making them with her own hands.

These early experiences were not isolated hobbies—they were the foundation of an engineering mindset. Her fascination with how things moved, fit, and functioned was not only natural but also deeply persistent. Each small act of construction was both a creative outlet and practical training, quietly building the skills that would later define her career.

Academic Pursuits

Upinder's academic journey was both strategic and driven by curiosity. She began by enrolling in a Bachelor's program in Mechanical and Automation Engineering, a decision shaped not just by her childhood fascination with machines and motion but also by her early realization that true innovation lies at the intersection of disciplines. While most students followed the curriculum, Upinder approached her education like an evolving project, constantly seeking ways to broaden and deepen her knowledge beyond what was prescribed.

In her undergraduate years, she quickly distinguished herself as someone who was not content with mastering just one domain. The foundational mechanical design principles, such as thermodynamics, fluid mechanics, and materials science, captivated her. But what excited her most was the potential to bring mechanical systems to life through automation and intelligent control. This realization pushed her to dive into areas typically outside the mechanical realm.

Recognizing that the future of engineering lay in autonomy and intelligent systems, she began supplementing her coursework with programming. She taught herself to code in languages like C++, Python, and MATLAB. The logic and design aspects of coding made it a joy.

Discovering Robotics

Upinder Kaur experienced a transformative shift in her academic focus in the final years of her undergraduate program. While mechanical design and automation had always intrigued her, it was during this phase that she discovered the world of robotics which seemed like a domain where all her interests seemed to converge. What drew her in was not just the complexity of robotic systems but the elegant way they merged the physical, digital, and cognitive realms. It was, in her eyes, the future of engineering, and she was determined to be a part of it.

She then decided to pursue a Master's in Robotics and Automation. In this program, she brought all her previous learning together, working on advanced projects that combined precision mechanics, real-time control systems, and intelligent algorithms. Each academic milestone she reached was not just about ticking boxes, but it was part of a deliberate effort to become fluent in the full language of robotics.

Unlike many students who confined their learning to lecture halls and labs, Upinder ventured far beyond the classroom. She read books she could find on robotics, control systems, artificial intelligence, and embedded programming. She also found online platforms for MOOCs and tutorials, especially during summer vacations to further improve her skillset. Through platforms like EdX and MIT OpenCourseWare, she taught herself the core concepts of kinematics, path planning, sensor integration, and machine learning, long before they appeared on any official syllabus.

Standing Alone

As the only woman in her mechanical engineering cohort, a class of nearly 60 students, Upinder often felt like she was charting new territory. The environment was intensely competitive and, at times, isolating. There were moments of doubt from others. Some professors questioned her choice of field, asking why she chose mechanical engineering—a question that seemed more common for her than for her male peers. These interactions felt more like skepticism than curiosity. One professor even asked how long she planned to stay in the program, as if anticipating she might change direction.

Rather than being discouraged by these experiences, Upinder found them motivating. She channeled every doubt into determination. Instead of retreating, she studied harder, asked sharper questions in class, and volunteered for the most technically challenging components in group work, such as CAD design, thermal calculations, and complex simulations. She knew that to succeed in this environment, she could not just be good. She had to be exceptional.

And exceptional she was. Upinder consistently earned top marks across subjects, often outperforming her peers. Her academic transcripts became clear demonstrations of capability. Over time, the tone shifted. Professors who once doubted her began to acknowledge her intellect, often through the increased responsibilities they entrusted to her. Her classmates grew more accepting as she aced thermodynamics exams and led successful final-year projects.

Most importantly, this period taught her the value of persistence and self-advocacy. She had chosen a path that was neither easy nor always welcoming, and she walked it with grit and grace.

Professional Beginnings

When it came time to enter the workforce, Upinder encountered some interesting contrasts in how she was perceived. During campus placement drives and job interviews, she noticed that male candidates were immediately quizzed on their technical knowledge. They were asked to solve problems, draw schematics, or explain how they would optimize engine performance or troubleshoot a system. The focus was clearly on technical competence.

Upinder's interviews often took a different path. Before reaching the technical round, she frequently found herself answering questions about her career motivations:

> "Why did you choose mechanical engineering?"
> "Are you sure this is the right field for you?"
> "Do you see yourself continuing in this industry long-term?"

These questions seemed to focus more on her commitment than her capabilities, as if her academic record and project work needed additional validation.

Upinder approached these conversations with patience and clarity. When finally allowed to discuss her technical work, she impressed interviewers with her deep understanding of control systems, her hands-on projects, and her proficiency in programming and electronics. Her ability to bridge mechanical hardware with embedded code was valuable.

Her persistence paid off when she landed a role at General Motors Technical Center India, a company known for its rigorous technical standards and cutting-edge automotive engineering. There, she found herself in the electrifications and controls division as a senior software engineer. She quickly adapted to the fast-paced, high-stakes environment, where precision and timing were everything.

Working with real-time operating systems and microcontroller-based applications, she dove into tasks like testing firmware for vehicle control units and testing systems in simulated and physical testbeds. She contributed directly to the car's brain: systems that governed safety, efficiency, and performance.

Not only did she keep pace with her teams but often led it. Within months, she was trusted with greater responsibility and began managing components of key projects. Her colleagues came to rely on her accuracy, her systems thinking, and her calm approach to debugging under pressure.

Joining Purdue

Her transition to research took shape at Purdue University, where she joined a lab within the Engineering and Technology Department known for its cross-disciplinary focus. This environment brought together projects spanning surgical robotics, precision agriculture, and intelligent systems. It was a whole new world of innovation that aligned perfectly with her broad and self-developed skillset. Having built a strong foundation in mechanical systems, programming, and electronics, she was well-prepared to take on the challenges of this diverse research space.

Rather than being confined to a single specialization, Upinder thrived in the lab's interdisciplinary nature. Each project allowed her to apply and extend different areas of her expertise, reinforcing her decision to move from industry into academia. In this dynamic setting, she began to see how her varied experiences could come together to support meaningful, real-world applications of robotics and automation.

Early Breakthroughs

During her time at Purdue, she took on a unique and unexpected challenge: applying robotics to animal health. This was far outside her previous experience. She had not worked in agriculture before, and she openly admitted that she was initially intimidated by working with large animals like cows. Yet, when the opportunity arose to work on a project in precision animal agriculture, she did not shy away. Instead, she embraced the challenge, recognizing that real innovation often happens when you step outside your comfort zone.

The work required her to learn not just new technologies but entirely new contexts. She had to understand how animals behave, design systems that could monitor their health without causing stress, and bring robotic tools into unpredictable and dynamic environments. Through close collaboration with domain experts and a willingness to start from scratch, she developed a cyber-physical system with edge robots for monitoring their health.

Her contribution was not just technical—it was foundational. She developed datasets, models, and algorithms for sensors and robots to work in the stomach of a cow. It was a rare blend of robotics, perception, and systems engineering applied to a problem with real-world impact.

This project became a defining moment in her research career, not just because of the novelty of the work, but because it demonstrated her ability to approach unfamiliar problems with humility, persistence, and a broad, adaptable skillset.

The Birth of a Lab

Now leading her own lab at Purdue University, Upinder Kaur focuses on three primary research pillars: entity-aware interaction, neuromorphic computing, and soft robotic design. Her work is paving the way for the next generation of robotics that can perceive and interact in real-time, low-power environments such as farms and forests.

Animal–Robot Interaction Upinder's research in entity-aware interaction is inspired by the natural behaviors of shepherd dogs, which can maneuver and control herds of cattle without a common language. Her goal is to develop robotic systems that can similarly engage with animals, other robots, and humans, interpreting their movements and behaviors to interact meaningfully. This involves creating algorithms and systems that allow robots to understand and respond to the entity's dynamic and unpredictable behavior. This research has significant implications for addressing labor shortages in agriculture, enhancing farm safety, and improving the precision and customization of animal care.

Neuromorphic Computing Upinder is exploring how to mimic the efficient processing capabilities of biological brains in robotic systems. Neuromorphic computing involves designing hardware and software that emulate the neural structures and functions of the brain, enabling robots to process sensory information rapidly and with minimal power consumption. This approach is particularly valuable for deploying robots in austere environments where resources such as power and communication bandwidth are limited. By integrating novel sensors with neuromorphic computing, Upinder's research aims to create robots that can operate effectively in real time, efficiently adapting to their surroundings. This technology is crucial for agriculture, forestry, and wildlife-monitoring applications, where robots must perform complex tasks under challenging conditions.

Soft Robot Design Upinder's lab is also actively building new robotic structures, including underwater robots, embedded intelligent skins, and ground robots adaptive to various terrains. This is geared toward exploring new avenues and environments, while advancing the forefront of robotics design.

Through these pillars, Upinder's lab is at the forefront of developing innovative robotic solutions that are technologically advanced but also practical and sustainable for real-world applications. Her work sets new standards for how robots can interact with living beings and their environments, making significant strides toward a future where autonomous systems play a vital role in enhancing agricultural productivity and environmental conservation.

Building Balance

Despite the demands of leading a research lab, Upinder maintains several practices that keep her grounded and creative: painting, gardening, and lifting weights. These activities serve different but equally important functions in her life.

Painting Upinder's passion for painting dates back to her childhood, where she found solace and expression through art. Today, painting provides a mental reset from her technical work, offering a different kind of intellectual stimulation that complements her scientific endeavors. This creative practice helps her maintain cognitive flexibility and often sparks new ways of thinking about research problems.

Gardening Growing vegetables and plants connects Upinder to natural systems in a hands-on way. The physical activity involved in gardening and the satisfaction of nurturing plants from seed to harvest offers a calming counterpoint to her research environment. Gardening also reinforces lessons about patience, observation, and adaptation, qualities that serve her well in research.

Fitness Physical fitness through cardio, strength training, and yoga has become an essential part of Upinder's routine. It has become essential for maintaining her energy and focus. Regular exercise helps her stay physically and mentally sharp, and the discipline required translates well to the persistence needed in research. The gym provides a space for reflection and stress relief that enhances her overall productivity.

Together, these diverse activities and the supportive institutional culture create a sustainable approach to her career. They help her maintain perspective, restore creativity, and prevent burnout, enabling her to excel in her research and contribute meaningfully to her field.

Advice for the Future

To young women aspiring to enter STEM fields, Upinder offers advice based on her own experiences and challenges. She says find what motivates you and what is interesting to you. Once you find that area or domain, be prepared, be persistent, and find good mentors.

Find Your True Interest The foundation of everything is genuine curiosity and passion. Upinder emphasizes that you cannot sustain yourself through the inevitable challenges of STEM without a deep, personal connection to your work. For her, it was the childhood dream of space exploration that led to mechanical engineering, which evolved into robotics when she discovered how perfectly it combined all her interests. Look for the intersection of what excites you intellectually and what you are naturally good at. This intrinsic motivation will carry you through difficult moments when external encouragement may be lacking.

Be Prepared Once you have found your passion, thorough preparation becomes your competitive advantage. Upinder learned early that she needed to be exceptionally well-prepared for every opportunity and challenge. This means not only mastering the required coursework but also going beyond and teaching yourself additional programming languages, taking on challenging projects, and anticipating what skills you will need next. Preparation also means being ready for unexpected opportunities. When the chance came to work on agricultural robotics, her self-taught programming skills and hands-on project experience made her ready to tackle something completely new. Being overprepared gives you confidence and demonstrates your commitment to your field.

Be Persistent Persistence is not just about pushing through obstacles—it is about maintaining your vision even when others cannot see it yet. Upinder's journey required her to stay focused on her goals despite the moments of doubt from others and the natural challenges of breaking into a new territory. Persistence means continuing to pursue excellence, continuing to learn, and continuing to contribute even when recognition is slow to come. It also means being willing to start over when necessary, like when she transitioned from industry to academia, or when she tackled animal robotics despite having no agricultural background. Each setback becomes data for your next approach.

Find Good Mentors While much of Upinder's early journey was self-directed, the importance of mentorship became clear as her career progressed. Good mentors are not just senior people in your field, but they are individuals who believe in your potential and are willing to invest in your growth. This includes professors who push you to think bigger, colleagues who collaborate generously, and industry professionals who share honest advice about career paths. Sometimes mentorship comes from unexpected places—domain experts in other fields, peers who become collaborators, or even junior colleagues who bring fresh perspectives. The key is to actively seek out people who support your vision and can help you navigate the unwritten rules of your field.

Upinder's central message is that success in STEM starts with authentic passion but requires the practical skills of preparation, the resilience of persistence, and the wisdom that comes from learning from others. When these elements work together, they create not just career success but the ability to make meaningful contributions to your field and open doors for the next generation.

Building Dreams: The Hands-On Journey of Devon Pessler in Engineering Technology

Devon Pessler

Introduction

Devon Pessler's journey into engineering technology is a testament to passion, perseverance, and the pursuit of hands-on learning. From childhood dreams of fantastical architectural designs to becoming a professor at Purdue University, Devon's story is filled with pivotal moments, challenges, and inspiring mentors.

Devon's fascination with building and creating began at a young age with Lego bricks. Her imaginative designs, including pizza rooms and backyard baseball fields, sparked an initial interest in architecture. However, as she progressed through high school, Devon's path shifted toward engineering technology, influenced by the Project Lead the Way courses and her first CAD model of a motor mount.

Devon's educational journey took her from California to Purdue University, driven by a desire for a strong engineering program and proximity to family. Her exploratory start at Purdue led her to the Mechanical Engineering Technology (MET) program, where the hands-on approach resonated deeply with her interests. Devon's involvement in CAD modeling and 3D printing during high school laid a solid foundation for her future studies.

Throughout her academic career, Devon faced significant challenges that tested her determination. One notable moment was turning down a CTO position at a startup to focus on her master's project. Additionally, the personal struggle with her mother during the final years of her PhD required immense resilience. Despite these obstacles, Devon's commitment to her goals and the support of her mentors helped her persevere.

D. Pessler (✉)
Purdue University, West Lafayette, IN, USA
e-mail: dpessler@purdue.edu

A. Lucietto (ed.), *Innovating for Tomorrow: Engineering Technology*, Women in Engineering and Science, https://doi.org/10.1007/978-3-032-07645-8_14

Mentorship played a crucial role in Devon's journey. Mark French, her advisor, provided guidance through academic struggles and career decisions. Devon's father served as a steadfast support system, offering wisdom and encouragement. Paul McPherson, another mentor, helped Devon navigate the transition from student to professor, providing industry contacts and leadership opportunities.

Devon's professional role as a visiting professor at Purdue University involves teaching hands-on courses like ENGT 182 and MET 102. Her passion for CAD modeling and practical applications is evident in her teaching approach. Devon's involvement in the Society of Manufacturing Engineers (SME) and the Technical Assistance Program further highlights her commitment to hands-on learning and industry collaboration.

Balancing work and life has been a continuous effort for Devon. While the first semester as a professor was overwhelmingly work-focused, she has since made strides toward a healthier balance. Activities like visiting state parks, working out, and spending time with her dog contribute to her well-being. Devon's efforts to engage in community activities and maintain social connections reflect her dedication to a balanced lifestyle.

Devon aspires to secure a permanent position at Purdue University and continue her teaching career. She envisions a future where women play a more significant role in engineering technology, driven by increasing interest in high-tech and hands-on fields. Devon's deep appreciation for initiatives like the Women in Technology organization and SWE's participation in Grand Prix exemplifies her commitment to empowering women in the field.

Devon Pessler's journey is a remarkable example of how passion, mentorship, and resilience can shape a successful career in engineering technology. Her story serves as an inspiration to aspiring engineers, highlighting the importance of hands-on learning, overcoming challenges, and seeking support from mentors. As Devon continues to contribute to the field, her impact on students and the broader engineering community will undoubtedly grow.

Early Inspirations

Devon's fascination with building and creating began at a young age with Lego bricks. These colorful bricks were more than just toys; they were the foundation of her creativity and imagination. As a child, Devon spent countless hours constructing elaborate structures, each one more ambitious than the last. Her designs were not just simple houses or buildings; they were imaginative creations that included pizza rooms and backyard baseball fields. These fantastical renditions were a reflection of her dreams and aspirations, and they sparked an initial interest in architecture.

Devon's early experiences with Lego bricks were not just about play; they were about problem-solving and innovation. She enjoyed the challenge of figuring out how to make her designs work, and this problem-solving mindset would later become a crucial part of her engineering career. The joy she found in building and creating with Lego bricks laid the groundwork for her future studies and career in engineering technology.

As Devon grew older, her interests began to evolve. While architecture initially seemed like the natural path for someone with creative talents, she soon discovered a new passion: engineering. This shift was influenced by her participation in the Project Lead the Way courses during high school. These courses introduced her to the world of engineering and provided her with hands-on experience in designing and building. It was during this time that Devon created her first CAD model of a motor mount for a 1929 Model A Roadster frame to connect to a 2006 Ford Mustang V6 engine, a project that would become a pivotal moment in her educational journey.

The Project Lead the Way courses were instrumental in shaping Devon's career path. They provided her with the opportunity to explore different aspects of engineering and to develop practical skills that would be essential in her future studies. The hands-on nature of these courses resonated with Devon, and she began to see the advantages of pursuing a career in mechanical engineering over architecture. The experience of creating a CAD model and seeing it come to life through 3D printing was a turning point for Devon, solidifying her decision to pursue engineering technology.

Devon's early inspirations were not limited to her experiences with Lego bricks and high school courses. Her family also played a significant role in shaping her career path. Growing up in a family with a strong sports background, Devon was encouraged to pursue her interests and to strive for excellence in everything she did. Her father's suggestion to consider Purdue University, known for its strong engineering program and excellent basketball team, was a key factor in Devon's decision to attend the university. The combination of family support, hands-on experiences, and a passion for building and creating set the stage for Devon's successful career in engineering technology.

Educational Journey

Devon's educational journey is marked by strategic decisions and transformative experiences that guided her toward a career in engineering technology. Her path began in California, where she explored various UC and Cal State universities. However, her family's connection to Cincinnati and her father's recommendation to consider Purdue University ultimately led her to West Lafayette, Indiana.

Upon arriving at Purdue, Devon was initially uncertain about which engineering discipline to pursue. She enrolled as an exploratory student, allowing herself to explore both Mechanical Engineering Technology (MET) and Mechanical Engineering (ME). This decision proved pivotal, as it enabled Devon to experience the unique aspects of each program and make an informed choice about her future.

Devon's first exposure to the MET program came through the MET 162 course taught by Professor Hillsman. The course focused on analytical and computational problem-solving techniques, one being GFSA formatting, and provided Devon with a hands-on introduction to engineering technology. Despite the challenging nature of the course, Devon found herself drawn to the practical, hands-on approach that MET offered. The experience solidified her decision to pursue a degree in Mechanical Engineering Technology.

Devon's educational journey also included significant extracurricular involvement. As a thrower on the track team, she applied engineering principles to her training and performance, demonstrating the interdisciplinary nature of her interests. The unique combination of athletics and engineering gave Devon a holistic perspective on problem-solving and innovation.

In her sophomore year, Devon learned about the 3 + 2 program, an accelerated master's program that would allow her to earn her bachelor's and master's degrees in 5 years. Intrigued by the opportunity to further her education, Devon approached Professor Duane Dunlap to learn more about the program. When Devon expressed interest in the accelerated master's program, Dunlap challenged her to articulate her reasons for wanting to pursue further education. The question prompted Devon to reflect deeply on her goals and motivations. Despite initial uncertainty about her postgraduation plans, Devon's desire to explore academic research and deepen her knowledge led her to apply for the program.

A significant project marked Devon's master's studies: developing a wearable medical device, specifically a head brace for a patient in the Netherlands. The international capstone project provided Devon with invaluable experience in CAD modeling, material selection, and 3D printing (Pessler, 2019). Her dedication to the project was evident in her decision to continue working on it for her master's thesis despite her challenges and setbacks.

The transition from master's to PhD was driven by Devon's desire to explore the intersection of materials and engineering further. Her PhD research focused on the material selection for acoustic guitar soundboards, a project that combined her interests in materials science and music. Devon's hands-on approach to research involved extensive testing and collaboration with industry partners, including Gibson (Pessler, 2024).

In addition to her traditional graduate education, Devon was a student consultant for the Technical Assistance Program (TAP). The program allowed her to collaborate with local businesses and provide consulting services to improve their manufacturing processes. As a student consultant in the TAP, Devon worked with companies to identify bottlenecks and propose solutions that enhance efficiency and productivity. Her hands-on problem-solving approach and ability to apply engineering principles in practical settings made her a valuable asset to the program and rounded out her education with industry experiences that no other assistantship offered. Through TAP, Devon contributed to the local manufacturing community and gained insights that enriched her education. Devon's efforts were also recognized as she received the "Industrial Engagement Excellence Award" at the end of her 7-year tenure with TAP.

Throughout her PhD studies, Devon faced significant personal challenges, including a strained relationship with her mother. Despite these obstacles, she remained committed to her academic goals, supported by her mentors and her father. Devon's resilience and determination enabled her to complete her dissertation and contribute valuable insights to the field of materials science.

Devon's educational journey culminated in her transition to a professional role as a visiting professor at Purdue University. Her passion for hands-on learning and practical applications continued to drive her teaching and industry engagement. Devon's involvement in the Society of Manufacturing Engineers (SME) and the

Technical Assistance Program further demonstrated her commitment to bridging the gap between academia and industry.

Challenges and Determination

Devon's journey was not without its share of challenges, each of which tested her determination and resilience. One of the most significant challenges came during her master's studies when she was considered for the position of Chief Technology Officer (CTO) at a startup company called Presso. The opportunity was a tempting prospect, promising a dynamic role in a burgeoning company founded by Purdue alumni. However, Devon was deeply committed to her master's project, which involved the development of a wearable medical device. She faced a difficult decision: to pursue the exciting opportunity at Presso or to remain loyal to her academic project.

Ultimately, Devon chose to stay with her master's project, demonstrating a profound sense of loyalty and dedication. This decision was not easy, as it meant turning down a potentially life-changing opportunity. However, Devon's commitment to her academic goals and her belief in the importance of her project guided her choice. This moment highlighted Devon's determination to see her projects through to completion, even when faced with significant temptations.

Another major challenge arose during the final stages of Devon's PhD. As she approached the defense of her dissertation, she experienced moments of doubt and uncertainty. The pressure of completing her PhD in a condensed timeframe, coupled with the personal struggles she was facing, made the period particularly challenging. Devon's father played a crucial role in helping her navigate this difficult time. Her father's advice to consider the broader perspective and the value of the journey rather than just the end goal gave Devon the encouragement she needed to persevere.

Devon's relationship with her mother posed a significant challenge during her academic journey. The strained relationship created emotional turmoil that affected her focus and well-being. Despite these personal struggles, Devon remained committed to her academic goals. She made the difficult decision to distance herself from her mother, recognizing that it was necessary for her mental health and academic success. The decision required immense strength and resilience, as it involved letting go of a significant part of her life.

Throughout these challenges, Devon's mentors provided invaluable support. Mark French, her advisor, offered guidance and encouragement, helping Devon navigate the complexities of her academic projects and personal struggles. Paul McPherson, another mentor, provided practical advice and industry connections, helping Devon transition from student to professor. These mentors played a crucial role in helping Devon overcome obstacles and stay focused on her goals.

Devon's determination was also evident in her approach to research. Her PhD project on material selection for acoustic guitar soundboards involved extensive testing and collaboration with industry partners. Despite facing setbacks and challenges in her studies, Devon remained committed to finding solutions and advancing her knowledge. Her hands-on research approach and her willingness to learn from failures demonstrated her resilience and dedication to her field.

In summary, Devon's journey was marked by significant challenges that tested her determination and resilience. Her ability to navigate these obstacles, supported by her mentors and her own inner strength, enabled her to achieve her academic goals and contribute valuable insights to the field of engineering technology. Devon's story is a testament to the power of perseverance and the importance of staying committed to one's goals, even in the face of adversity.

Mentorship and Support

Mentorship and support have been pivotal in Devon Pessler's journey, providing her with guidance, encouragement, and practical advice at crucial moments. Three key figures stand out in Devon's narrative: Mark French, her father, and Paul McPherson.

Mark French As Devon's academic advisor, Mark French played a central role in shaping her educational and professional trajectory. French's mentorship extended beyond academic guidance; he gave Devon a broader perspective on the field of engineering technology and the opportunities it offered. French's approach was holistic, considering not only Devon's academic performance but also her personal development and career aspirations.

French's influence began early in Devon's academic journey. Devon had Mark French as a professor during her junior year when she took dynamics. She saw the way French used real-world examples in his lectures and made his course's final project: building either a trebuchet or an air cannon. This showed Devon that engineering technology was truly where she belonged as this was a perfect example of how her need for hands-on learning was met with a course's curriculum.

Throughout Devon's master's and PhD studies, French provided invaluable support and guidance. He helped Devon navigate the complexities of her research projects, offering valuable advice on material selection, material testing, and writing a thesis. French's expertise in these areas was instrumental in Devon's success, particularly in developing the wearable medical device and the acoustic guitar soundboard metric.

French also played a crucial role in helping Devon manage the challenges she faced during her PhD. When Devon struggled with the pressure of completing her dissertation and the personal turmoil caused by her strained relationship with her mother, French provided a steady source of encouragement and support. His mentorship helped Devon stay focused on her goals and maintain her determination in adversity.

Devon's Father Devon's father, Daryl Pessler, was a constant support and encouragement throughout her academic journey. His influence was both practical and emotional, providing Devon with the guidance she needed to navigate the challenges of her studies and her personal struggles.

Devon's father's suggestion to consider Purdue University was a pivotal moment in Devon's educational journey. His recommendation was based on a combination of academic excellence and personal interests, highlighting the importance of finding a balance between professional goals and personal passions. The advice set Devon on a path that would lead to significant academic and professional achievements.

Throughout Devon's studies, her father provided practical advice and emotional support. When Devon had to choose between seriously considering the CTO position at Presso or staying committed to her master's project, her father helped her weigh the pros and cons, ultimately guiding her to stay loyal to her academic goals. His support was crucial in assisting Devon to make a decision aligned with her long-term aspirations.

During the final stages of Devon's PhD, her father's encouragement was significant. As Devon grappled with doubts and uncertainty about completing her dissertation, her father provided a broader perspective, reminding her of the value of the journey and the lessons learned along the way. The advice helped Devon maintain her determination and ultimately achieve her academic goals.

Paul McPherson Paul McPherson's mentorship was instrumental in Devon's transition from student to professor. McPherson provided practical advice and industry connections, helping Devon navigate the complexities of her new role and establish herself as a faculty member at Purdue University.

McPherson's influence began during Devon's graduate studies when he offered Devon opportunities to gain practical experience in assisting with his courses, joining and leading Purdue's SME student chapter, and industry collaboration. These experiences were invaluable in helping Devon develop the skills and confidence needed to succeed in her professional role. McPherson's mentorship was characterized by a hands-on approach, providing Devon with real-world opportunities to apply her knowledge and skills.

When Devon took over McPherson's course, MET 102, McPherson provided detailed guidance on the class's inner workings and the expectations for teaching and student engagement. This support was crucial in helping Devon navigate the transition to a faculty role and establish herself as an effective educator.

McPherson also played a key role in helping Devon build industry connections and establish herself in the professional community. His advice on networking and collaboration helped Devon develop valuable relationships with industry partners, enhancing her ability to bridge the gap between academia and industry.

In summary, the mentorship and support provided by Mark French, Devon's father, and Paul McPherson were pivotal in shaping Devon's educational and professional journey. Their guidance, encouragement, and practical advice helped Devon navigate her challenges, stay focused on her goals, and achieve significant academic and professional success. Devon's story is a testament to the importance of mentorship and its profound impact on one's career and personal development.

Professional Role

Devon's professional role as a visiting professor at Purdue University is a testament to her dedication to hands-on learning and practical applications in engineering technology. Her teaching responsibilities include courses such as ENGT 182 and MET 102, which emphasize the importance of experiential learning and real-world applications.

ENGT 182 The course "Gateway of Engineering Technology" is designed to introduce students to the fundamental principles of engineering technology. Devon's approach to teaching ENGT 182 is characterized by her emphasis on hands-on activities and practical problem-solving. She believes that students learn best when applying theoretical concepts to real-world scenarios. In her classes, Devon incorporates a variety of interactive projects and laboratory exercises that allow students to experiment with different technologies and techniques. This approach enhances students' understanding of engineering principles and fosters creativity and innovation.

MET 102 The "Product Design and Specifications" course is another key component of Devon's teaching portfolio. The course focuses on CAD modeling and the creation of detailed production specifications. Devon's passion for CAD modeling is evident in her teaching methods, which include comprehensive tutorials, hands-on projects, and collaborative design challenges. She encourages students to explore CAD software's capabilities and develop their skills in creating precise and functional designs. Devon's expertise in CAD modeling, combined with her practical approach to teaching, helps students gain confidence in their abilities and prepares them for careers in engineering technology.

Devon's involvement in the Society of Manufacturing Engineers (SME) further highlights her commitment to hands-on learning and industry collaboration. As a co-faculty advisor for the SME student chapter at Purdue, Devon plays a crucial role in guiding and mentoring students interested in manufacturing engineering. She helps organize workshops, seminars, and industry visits that give students valuable insights into the manufacturing sector. Devon's efforts have contributed to the success of the SME student chapter, which was recently awarded the inaugural "Student Chapter of the Year" award. The recognition underscores the impact of Devon's mentorship and her dedication to fostering a strong connection between academia and industry.

Devon's professional role at Purdue University is characterized by her dedication to hands-on learning, practical applications, and industry collaboration. Her passion for CAD modeling and her commitment to experiential education are evident in her teaching methods and involvement in SME. Devon's efforts to bridge the gap between academia and industry profoundly impact her students, preparing them for successful careers in engineering technology and fostering a culture of innovation and collaboration.

Balancing Work and Life

Balancing work and life has been a continuous effort for Devon. The first semester as a professor was overwhelmingly work-focused as she navigated the demands of teaching large classes and adapting to her new role. With 340 students in ENGT 182 and 50 students in MET 102, Devon found herself immersed in lesson planning, grading, and managing classroom dynamics. The sheer volume of work left little room for personal activities, and Devon focused primarily on establishing herself as an effective educator.

Despite the initial challenges, Devon recognized the importance of achieving a healthier work–life balance. She began to make deliberate efforts to incorporate activities that contributed to her well-being and personal fulfillment. One of the key strategies Devon employed was spending time outdoors, exploring the natural beauty of Indiana's state parks. These excursions provided a much-needed respite from the demands of academia, allowing Devon to recharge and reconnect with nature. The physical activity and fresh air were beneficial for her mental and physical health, helping her maintain a positive outlook and reduce stress.

In addition to visiting state parks, Devon committed to regular exercise. She joined a gym off-campus, where she engaged in various fitness activities and began to build a routine that included working out several times a week. Devon's commitment to physical fitness improved her overall health and allowed her to meet new people and develop social connections. Her interactions at the gym helped her feel more integrated into the community and provided a sense of camaraderie essential for her well-being.

Another critical aspect of Devon's work–life balance was her relationship with her dog, Riley. Caring for her pet required her to establish a routine that included daily walks and playtime. These activities were enjoyable and provided a structured break from work, ensuring that Devon took time to relax and unwind. The companionship of Riley offered emotional support and a sense of responsibility, contributing to Devon's overall happiness and stability.

Devon also made efforts to engage in community activities and maintain social connections. She recognized the importance of a supportive social network and sought opportunities to participate in local events and activities. Whether attending community gatherings, joining clubs, or volunteering, Devon's involvement in the community helped her build meaningful relationships and feel more connected to those around her. These social interactions were crucial for her mental health, providing a sense of belonging and reducing feelings of isolation.

Over time, Devon's efforts to achieve a balanced lifestyle began to pay off. She found that incorporating activities that nurtured her physical, mental, and emotional well-being allowed her to approach her work with renewed energy and focus. The balance between professional responsibilities and personal fulfillment enabled Devon to be more effective as a professor, as she brought a positive and engaged attitude to her teaching.

Devon's journey toward a balanced lifestyle is a testament to the importance of self-care and the deliberate effort required to maintain well-being in a demanding profession. Her experiences highlight the value of outdoor activities, physical fitness, pet companionship, and community engagement in achieving a healthy work–life balance. By prioritizing these aspects of her life, Devon has created a fulfilling and sustainable routine that supports her professional success and personal happiness.

Future Aspirations

Devon aspires to secure a permanent position at Purdue University and continue her teaching career. Her experiences as a visiting professor have solidified her passion for education and her commitment to fostering hands-on learning environments. Devon's goal is to transition from a visiting professor to a full-time faculty member, where she can further develop her teaching methods, contribute to curriculum development, and engage in long-term industry-related projects.

Devon envisions a future where women play a more significant role in engineering technology, driven by increasing interest in high-tech and hands-on fields. She is dedicated to creating an inclusive and supportive environment for female students and professionals in engineering technology. Devon's deep appreciation of initiatives like the Women in Technology organization and SWE's participation in the Grand Prix exemplifies her commitment to empowering women. These organizations provide valuable resources, mentorship, and networking opportunities for women, helping them navigate the challenges of a traditionally male-dominated industry.

Devon believes that increasing the representation of women in engineering technology is crucial for advancing the field. Diverse perspectives and experiences can lead to innovative solutions and a more dynamic workforce. To support her vision, Devon plans to actively participate in outreach programs encouraging young women to pursue careers in STEM. She aims to collaborate with local schools, community organizations, and industry partners to create workshops, seminars, and mentorship programs that inspire and support female students.

In addition to her efforts to empower women, Devon is passionate about advancing the field of engineering technology through research and industry collaboration. She plans to continue her involvement in the Society of Manufacturing Engineers (SME) and the Technical Assistance Program (TAP), where she can contribute to cutting-edge research and provide practical solutions to industry challenges.

Devon's research interests include material science, CAD modeling, and additive manufacturing, which have significant potential for innovation and impact.

Devon also aspires to expand her role in professional organizations and contribute to developing national competitions and conferences that showcase the talents and achievements of engineering technology students. By fostering a competitive and collaborative environment, Devon hopes to inspire students to push the boundaries of their knowledge and skills, ultimately leading to advancements in the field.

Conclusion

Devon Pessler's journey is a remarkable example of how passion, mentorship, and resilience can shape a successful career in engineering technology. Her story inspires aspiring engineers by highlighting the importance of hands-on learning, overcoming challenges, and seeking support from mentors. Devon's dedication to education, commitment to empowering women, and involvement in research and industry collaboration demonstrate her profound impact on the field.

As Devon continues contributing to the field, her impact on students and the broader engineering community will grow. Her efforts to create inclusive and supportive environments, advance research, and foster industry collaboration will pave the way for future generations of engineers. Devon's journey is a testament to the power of perseverance, the importance of mentorship, and the transformative potential of hands-on learning. Her story encourages aspiring engineers to pursue their passions, overcome obstacles, and make meaningful contributions to the world of engineering technology.

Part IV
Current and Former Students (All Levels)

Current and former students' voices are essential to understanding engineering technology education's full impact and evolution. Their lived experiences offer a window into the effectiveness of curricula, the accessibility of academic pathways, and the real-world relevance of what is taught in classrooms and labs. This module, *Current and Former Students in Engineering Technology*, centers on the reflections, challenges, and triumphs of individuals who have navigated the engineering technology landscape from the student perspective.

The contributors to this module are Diana Navarez MSET, Colleen McGuinness, Esq., Julia Rowe, MSET, Meher Taleyarkhan, MSET, Olivia Engvall, Mica Caddick MSML, and Selena Maria Nadjarian they represent a diverse cross-section of learners and professionals. Their stories span community colleges, polytechnic institutions, and research universities, and they reflect a wide range of academic journeys, career goals, and personal motivations. Together, they provide a rich and nuanced portrait of what it means to be a student in engineering technology today.

Some authors share their experiences as students and some pursuing an education despite various challenges, balancing education with family responsibilities, or transitioning from remote places near the Amazon forest. Others reflect on their time as first-generation college students, navigating unfamiliar academic systems while striving to build a better future. Still others describe their paths through undergraduate and graduate programs, highlighting the importance of mentorship, hands-on learning, and industry engagement.

Throughout these narratives, common themes emerge: the value of applied, project-based learning; the importance of supportive faculty and peer networks; and the transformative power of perseverance. These students speak candidly about the barriers they faced whether financial, cultural, or institutional and how they overcame them. They also offer insights into how engineering technology programs can better serve future students, particularly those from underrepresented or underserved backgrounds.

This module is not only a celebration of student resilience and achievement, but it is also a call to action. By listening to the voices of those who have experienced engineering technology education firsthand, educators, administrators, and policymakers can gain critical insights into improving access, equity, and outcomes. These stories remind us that behind every program and policy are individuals whose lives are shaped by the opportunities and limitations of the systems we build.

As you read the chapters in this module, consider the diverse pathways that lead students into engineering technology, the support structures that sustain them, and the aspirations that drive them forward. Their journeys are not only personal—they are foundational to the future of the field.

From the Heart of the Amazon to the Frontiers of Engineering Technology: Diana Narvaez's Inspiring Journey

Diana Narvaez

Early Beginnings in Colombia

A young Diana with her parents in Villagarzón, Putumayo, Colombia, on the day of her kindergarten graduation.

D. Narvaez (✉)
Purdue University, West Lafayette, IN, USA
e-mail: narvaez@purdue.edu

A. Lucietto (ed.), *Innovating for Tomorrow: Engineering Technology*, Women in Engineering and Science, https://doi.org/10.1007/978-3-032-07645-8_15

Diana Narvaez's story begins in Villagarzón, a small town in the Amazon region of Colombia. Growing up in a remote and closely knit community came with its challenges but also shaped her deep appreciation for nature and collective support.

Both of her parents were deeply involved in her education, encouraging her curiosity from early on. Her mother, a high school math teacher, nurtured her interest in mathematics, while her father provided steady encouragement and belief in her potential. This environment sparked a lasting interest in math and science, which gradually expanded into a fascination with physics and its real-world applications.

Pursuing Engineering

Encouraged by her high school professors and driven by a desire to create meaningful change, Diana chose to pursue a degree in engineering. Coming from a small town, she faced limited access to academic resources, but her determination never wavered. She prepared thoroughly for the entrance exam to the National University of Colombia, the country's leading public university, and was admitted to study electronic engineering. Once in the program, she discovered a strong interest in electrical systems as well, which led her to pursue a dual major in electronics and electrical engineering. In her final year, she was awarded a scholarship for a research internship at Purdue University, an experience that later led to a Research and teaching assistantship and the opportunity to continue her graduate studies in the United States.

Engineering for Social Impact

Diana's early academic work was closely tied to the needs of her community, where electricity was often unstable. During her undergraduate studies, she focused on non-interconnected zones in Colombia, exploring how solar panels and renewable energy systems could improve the quality of life. This experience shaped her understanding of how technology can directly serve people's needs. While her current research in the United States has shifted toward materials science, flexible electronics, and biosensors, the same underlying goal remains: to develop technologies that have real social impact and can improve lives in underserved regions.

Family Influence and Support

Diana grew up in a family where education was seen as a form of empowerment. Her parents were the first in their families to earn university degrees, an achievement that shaped their values and set the tone for Diana's upbringing. Though they did not pursue STEM careers, they nurtured her academic curiosity with patience and conviction, encouraging her to explore her interests freely and aim high. Their stories of resilience in the face of limited opportunities served as quiet lessons in perseverance, and their belief in the transformative power of knowledge left a lasting impression.

Her grandparents, although they did not complete high school, were equally present in her journey. They may not have fully understood the details of her academic path, but their pride never wavered. Her grandmothers celebrated each new milestone, regardless of its complexity, with heartfelt joy.

This steady, multigenerational support gave Diana both confidence and responsibility to pursue her goals with purpose. As she advanced in her studies, she carried with her not only her own aspirations but also the hopes of those who had made her path possible. That sense of shared progress continues to shape her commitment to education, mentorship, and meaningful contribution.

Commitment to Education

Diana's commitment to education took root early, shaped by a home where learning was encouraged and curiosity was welcomed. Over time, this drive evolved from personal ambition into a broader sense of purpose. Earning a PhD was never just a title, it became a way to expand the horizon of what was possible for others around her.

Her younger cousins often turned to her for guidance, and Diana embraced that role with intention. She shared her journey openly, not to highlight achievements, but to normalize struggle, persistence, and the value of asking questions. For her, being transparent about setbacks was as important as celebrating success.

This sense of responsibility was reinforced by the steady encouragement she received from her family. Their confidence in her potential gave her strength through moments of doubt and helped her stay focused on the long view that education has the power to shift entire trajectories.

Today, her pursuit of a PhD is part of a larger commitment to mentorship and representation. In classrooms and labs, she sees opportunities not just to teach or research but to reflect back to others, especially young women from underrepresented backgrounds, the message that they belong in these spaces too. Her story is one of individual determination, but it is also a contribution to something larger: a culture where knowledge is shared, voices are lifted, and doors are held open for those who come next.

Teaching and Giving Back

Growing up, Diana witnessed the quiet power of teaching through her mother, a high school math teacher dedicated to serving students in underserved communities. During school breaks, Diana often returned to her hometown and joined her mother in the classroom. There she shared her academic journey with her mother's students and assisted with lessons. Many of these students came from indigenous communities and faced complex social challenge, and Diana hoped that hearing her story might inspire them to imagine new possiblities for their futures.

As a teaching assistant during her PhD, Diana came to appreciate the influence educators have far beyond the syllabus. She saw how inclusive, empathetic teaching could empower students to find confidence in their voices and take ownership of their learning. This reinforced her aspiration to become a professor and to cultivate learning environments where every student feels seen and capable.

Diana remains committed to returning to her community whenever possible. She envisions leading workshops and mentoring sessions that support students from underrepresented backgrounds, especially those who, like her, may be the first in their families to pursue higher education. For Diana, teaching is not only about sharing knowledge—it is about building bridges, expanding access, and creating lasting change where it is needed most.

Overcoming Challenges

Diana's experience in engineering came with academic demands and also with the challenge of finding her place in a space where women were underrepresented. In her early semesters, she was often the only woman in labs and group meetings. Speaking up felt intimidating, and her voice was frequently overlooked in conversations. Rather than collaborative, those environments felt isolating.

Everything shifted when she met one of the few female professors in her department. Their conversations helped Diana recognize that doubt and discomfort were common experiences for women in STEM. That validation made a difference. It gave her language for what she was feeling and the reassurance that she was not alone.

As time went on, Diana began to claim space with greater confidence. She spoke up more in class, took the lead on group projects, and began mentoring others. Those moments, once filled with hesitation, became steps forward. The support she received, and the courage she built, helped transform her outlook on engineering and on herself.

Her story shows how community and encouragement can help shift what is possible. With the right support, even the most exclusive environments can start to feel like places where new voices belong.

Transition to the United States

Arriving in the United States, Diana quickly noticed the contrast with her previous academic experiences. The environment felt more supportive, and even though women were still underrepresented, there were more of them in the classroom. Just seeing other female students made a difference, it created a sense of visibility and quiet reassurance.

In many classes, women naturally gravitated toward the same table, forming a small circle of mutual support. That simple act of gathering became a reliable space for discussion and learning. Within that group, ideas flowed more easily, and the pressure to prove oneself faded. Shared experiences, even unspoken, helped build trust and connection.

This sense of community played a key role in helping Diana adjust. It eased the isolation she had often felt earlier in her academic journey and gave her the confidence to speak up, take initiative, and fully engage. The friendships she built during this time became a source of strength and a reminder of the value of collective support, especially in spaces where many still feel like outsiders.

Language Barrier

Adjusting to a new academic environment came with a challenge Diana had not fully anticipated: speaking and teaching in English. At first, it felt uncomfortable, and she often worried about mispronunciations or not finding the right words. Instead of letting that hold her back, she turned the classroom into a space for shared learning.

She invited her students, many of them international, to help her improve. They offered feedback and corrections, and in doing so, saw someone willing to learn alongside them. This approach built trust and encouraged participation, creating a space where language differences became part of the learning process rather than a barrier.

Continued Underrepresentation

Despite the progress made in recent years, Diana still keenly felt the underrepresentation of women in her field, especially during conferences where male voices often dominated the discourse. This persistent challenge not only highlighted the systemic barriers women faced but also deepened her resolve to foster a more inclusive environment. Diana recognized that mentorship and support for women in STEM were crucial for overcoming these obstacles, and she committed herself to creating spaces where others could feel seen and supported.

One of the most pivotal moments in her journey was her collaboration with Dr. Newell, a prominent female advisor whose work and dedication Diana greatly admired. Dr. Newell's mentorship extended beyond academic guidance; she offered Diana a sense of safety and validation in an environment that had often felt unwelcoming and intimidating. This positive experience contrasted sharply with Diana's previous encounters in male-dominated labs, where she had often struggled to find her voice and assert her ideas.

Under Dr. Newell's guidance, Diana not only grew as a scientist but developed a profound appreciation for the importance of representation. Inspired by her advisor's example, she now prioritizes inclusive teaching and mentorship in her own work. In her classrooms, she pays special attention to supporting women and international students, encouraging them to engage fully and share their perspectives. When she returns to her hometown, she visits local schools to share her story, offering guidance and encouragement, especially to young women who may be unsure if STEM is for them.

Ultimately, Diana's journey is about more than personal achievement; it reflects a commitment to collective progress. Her experience with Dr. Newell reinforced her belief in the power of mentorship, not only to support individual growth but to open doors for others and foster change within a community.

Mentorship and Support

Diana's path has been shaped by mentors who created spaces where she felt respected and supported. During her master's program, she worked with Dr. Leon-Salas, a Peruvian professor actively involved in minority-focused programs. His lab brought together students from diverse backgrounds—including Latinos, Asians, Europeans, and Africans—and offered a level of inclusion and encouragement that stood in stark contrast to the challenges Diana had faced in earlier academic environments. That experience helped her gain confidence and recognized the importance of mentorship rooted in trust and belonging.

Later, during her PhD, Diana continued to grow under the guidance of Dr. Newell. She appreciated the support and openness her advisor offered, which allowed her to pursue her research with independence while feeling valued and seen. Dr. Newell's example, grounded in integrity and thoughtful leadership, has also been a quiet source of inspiration. Both mentors, in their own ways, played a vital role in shaping the academic environment Diana now strives to create for others.

Admiration for Female Role Models

Diana's earliest role model was her mother, a full-time teacher who balanced work and family with determination and care. As a student in Colombia, she was further inspired by two women professors whose presence in engineering classrooms helped normalize women's leadership in technical spaces.

Later, during her PhD, Dr. Newell became a key figure in her academic path. Her openness about the realities of combining motherhood with a research career gave Diana a realistic and empowering vision of the future. In many ways, Dr. Newell became a role model not just for her success but also for her honesty and balance. Her example helped Diana believe that it was possible to build a life where career and family could coexist, even if imperfectly—and that embracing that complexity could be a strength, not a limitation.

Encouraging Future Generations

Diana Narvaez (far right) with her students and her faculty advisor (second from the right) during an engineering competition.

Diana's experience as a student continues to shape how she teaches. Having once felt hesitant to ask questions, she understands how uncertainty can silence participation. This drives her commitment to creating classrooms where curiosity is welcomed and every voice matters.

She encourages open dialogue and reminds her students—especially women—that asking questions is part of learning. By fostering a culture of inquiry, she helps build confidence and engagement.

What matters most to her is the trust and community that emerge in these spaces. Through guidance, honesty, and persistence, Diana shows her students that growth comes through effort and that the classroom can be a powerful place to begin believing in themselves.

Current Work and Research

In addition to her role as a teaching assistant, Diana teaches several undergraduate courses, including ECET 337, ENGT 182, and the lab sequence 339–349. She draws on her interdisciplinary background, including time in the Spanish department, to create engaging and inclusive lab experiences. Her research centers on sensors, actuators, and flexible electronics, integrating concepts from electrical engineering, materials, science, and mechanics to develop emerging technologies. With current applications in tunnel environments, work that blends her foundations in electrical engineering with challenges in materials science and mechanics.

Diana approaches each new research challenge with curiosity and persistence, viewing technical complexity as a space for growth. Her ability to adapt across disciplines has allowed her to explore innovative paths within engineering.

As the first woman PhD student in Dr. Newell's lab, Diana has appreciated the supportive and respectful mentorship she has received. Her encouragement has contributed to a positive research environment where exploration and self-direction are valued, reinforcing Diana's confidence as she continues to shape her trajectory in the field.

Perspective as a Woman in Engineering Technology

Diana's journey in engineering technology is deeply connected to her commitment to inclusivity and support for others. Her experience as a woman in a predominantly male field gives her insights into the barriers that many underrepresented students face in STEM. This perspective shapes the way she interacts with her peers and students and informs the environment she works to build.

In her role as a TA, Diana fosters a classroom atmosphere that encourages participation and mutual respect. Aware that many students arrive with experiences that have challenged their confidence, she takes time to ensure that every voice feels valued. By checking in regularly and staying accessible, she creates a space where students feel a genuine sense of belonging.

Her memories of feeling isolated during her undergraduate years continue to motivate her. Those early moments of uncertainty now fuel her drive to teach with empathy and intention. She encourages group discussions, adapts to different learning styles, and opens space for honest conversations.

Diana is especially committed to mentoring students from marginalized backgrounds, offering both guidance and a relatable presence. She shares her own story when it can help others move forward and aims to be someone her students can look to not just for answers but also for encouragement.

She believes that creating inclusive spaces strengthens both individuals and the field as a whole. When students feel seen and supported, they are more likely to take risks, grow, and contribute meaningfully to the future of engineering technology.

Balancing Life and Work

Balancing life and work has been a constant challenge for Diana, especially without family nearby. Her closest support system consists of fellow PhD students, many of whom are navigating similar pressures, which makes it hard to fully disconnect from academic life. She has been pleasantly surprised by the wide range of clubs and activities available in the United States, options that were less common during her studies in Colombia, and she has made an effort to explore them as a way to reconnect with herself outside the lab.

During her master's program, Diana often felt isolated and fully consumed by research. That experience led to frustration and moments of imposter syndrome. To cope, she began building small routines into her day, like attending elective classes and going to the gym, which helped her regain motivation and balance. Winters continue to be especially tough, affecting her mood and energy, but she actively looks for ways to stay engaged, prioritizing her mental health alongside her academic goals.

Encouraging Students to Find Balance

Diana understands the importance of maintaining a balance between academic work and personal life. She regularly checks in with her students, asking about their weekend plans and encouraging them to engage in fun activities. Diana believes that having a life outside the lab is crucial for motivation and mental health. She shares stories about the importance of balance and the impact it can have on one's overall well-being.

Diana's approach to mentoring reflects her belief in the importance of balance. She encourages her students to find extracurricular activities and hobbies that they enjoy. By fostering a supportive and balanced environment, Diana aims to help her students navigate the challenges of academic life while maintaining their mental and emotional health.

Cultural and Gender Differences in the Workplace

Diana's international experiences have shaped her understanding of how gender is perceived in academic and professional spaces. During a summer research stay in Saudi Arabia, she encountered a cultural environment that felt markedly restrictive. Although no one made overtly chauvinistic remarks, the underlying societal expectations and limitations placed on women created a subtle but constant sense of constraint.

In Colombia, the challenges were more direct. Diana often heard comments, sometimes from friends or extended family, questioning her place in engineering, reinforcing the notion that it was a career for men. These remarks made her doubt herself at times, but the steady encouragement from her parents helped her push through and stay focused on her goals.

Her experience in the United States has been notably different. Here, she has found a more supportive and equitable environment, where she feels free to pursue her work without the added burden of gender-based judgment. Mistakes are seen as part of learning, not as a reflection of her worth, and this has allowed her to grow with confidence.

Reflecting on her path from the Amazon to academic labs around the world, Diana feels a deep responsibility to inspire young girls—especially those in Colombia, the United States, and Saudi Arabia, who may be navigating similar questions about their place in science and technology.

Advice for a Little Girl from Colombia

To a little girl in Colombia who dreams of becoming an engineer, Diana would say: *You don't have to feel ready to begin, just willing*. When she started her bachelor's degree, she constantly felt out of place, unsure if she belonged in engineering. But she learned to embrace that discomfort as a sign of growth: "People around you might say that engineering isn't for women. It can make you question your path. But trust yourself. Learning is hard for everyone, no matter how smart they are. What matters is showing up, asking questions, and staying curious. Let your determination speak louder than your doubts."

Advice for a Little Girl from the United States

To a girl in the United States who is thinking about engineering, Diana would say: ***Yes, you can absolutely do this***. A degree in STEM gives you tools that can change the world, and your own life too. Don't let fear of difficulty stop you. Grad school, careers in tech, and leadership roles may seem intimidating, but they are reachable." Diana often tells her students: "apply, try, ask for feedback. If you need support, there are people, like her, who want to help. You don't need to be perfect. You just need to start. The world needs your ideas and your voice."

Advice for a Little Girl from Saudi Arabia

To a girl in Saudi Arabia, Diana would offer this: *Living outside your comfort zone may be the hardest thing you'll ever do, but also the most powerful.* When she spent a summer in Saudi Arabia, Diana saw how cultural expectations shaped women's experiences in subtle but deep ways. Even without direct discrimination, the pressure was real. But she also saw how asking questions and building bridges between cultures can change perspectives. Moving from Colombia to the United States, and later living in Saudi Arabia, taught her that "you don't need to fit a mold. Your experience, your strength, and your curiosity are enough to lead you forward."

Importance of Exploring Beyond Academia

As a PhD student, Diana recognizes that learning does not stop at coursework and lab work. She values stepping outside her immediate research focus to stay informed about related fields. Reading, attending conferences, and talking with other researchers have helped her see new connections and reminded her she is not alone in facing academic challenges. These interactions not only bring fresh insights but also foster a sense of community and shared purpose.

Her collaboration with Eli Lilly and Toyota gave her a firsthand look at how research functions outside of academia and in industry. Unlike academia, where timelines can stretch and failure is part of the process, she saw the urgency and financial consequences of every delay. The pressure to deliver results was intense, and daily failures had real costs. This experience helped her appreciate the value of understanding both academic and industrial perspectives and the importance of bridging those worlds through communication, collaboration, and adaptability.

Core Skills and Qualities for Success in Engineering Technology

Through her academic and professional journey, Diana has identified several core qualities that she considers essential for success in engineering technology. **Adaptability** stands out; living and working in different countries have taught her the value of being flexible and open to new ideas. **Collaboration** is equally important, as working with others and sharing knowledge often lead to more effective solutions and deeper understanding.

Persistence has been a defining trait in Diana's path. She encourages young women to trust themselves, especially when stepping outside their comfort zones. **Curiosity** also plays a key role; a genuine interest in learning and exploring new areas can greatly enhance problem-solving and innovation.

Beyond technical skills, Diana emphasizes **empathy**, the ability to support and understand others, particularly those who may feel underrepresented. Her commitment to inclusion and mentorship is rooted in this value. She also highlights the need for **real-world awareness**. Her work with Eli Lilly gave her insights into the financial and operational pressures in industry, reinforcing the importance of staying connected to the practical implications of research.

She encourages students to stay informed, read beyond their specialization, and develop both their technical and interpersonal strengths. In her view, these qualities truly empower engineers to thrive and lead.

Impactful Research Projects

Diana is currently focused on flexible electronics, sensors and emerging soft electronic systmes Her work explores how conductive mateiials and flexible structures can be integrated to create responsive devices and with applications in healthcare ad human centered technologies. One area of interest includes the development off wearable systmes designed to suppoort hadn rehabilitation in elderly individuals and patients recoring from injuries. For Diana, the potential to improve quality of life through engineering innovatino is what makess this work expecially meaningful.

Over time her interest in socially impactful research has continued to grow. Earlier in her studies, she sometimes questioned how tradiional elecctronic research translated into real world benefits. Focusing on technologies that interact directly with people helped her see new possibiities for engineering to address human needs. Through this work she aims to contribute to technologies that are not only technically advanced, but also capbile of improving every day life.

Advice for Young Women in STEM

Diana encourages young women, especially those from Colombia, to trust themselves and step beyond their comfort zones. She knows from experience that learning is rarely easy, and self-doubt is common, but persistence makes all the difference. Engineering may feel intimidating at times, but growth happens when you lean into discomfort and keep moving forward.

To young women in the United States, she offers similar encouragement: "pursue your passion, and don't underestimate your ability to make an impact." Diana often shares her own path, including her transition into industry and her decision to pursue graduate school, even when it felt uncertain. She reminds students that "applying to grad school takes courage, but it's worth it. The key is to believe in your potential and take that first step."

Through her story, Diana hopes to show that success in STEM is not about being fearless—it is about staying curious, being resilient, and creating space for others along the way.

Real-World Pressure and Industry Experience

Diana's collaboration with Eli Lilly gave her a firsthand look at the pressure and pace of industry research. Working on data-driven projects with Dr. Newell, she was struck by how different the environment felt from academia. One manager explained the financial consequences of delays; daily failures could mean tens of thousands of dollars lost. The urgency was intense, and for Diana, it was eye-opening.

That experience reinforced the importance of exposing students to the realities of industry. Academic research can feel theoretical, but in the real world, time, money, and results are tightly linked. Diana now believes that STEM education should include honest discussions about funding, deadlines, and impact. Understanding these pressures, she argues, helps students become not just better researchers, but more adaptable and prepared professionals.

A Vision for the Future

Diana Narvaez celebrates her Master graduation at Purdue University

As Diana approaches the end of her PhD, her vision extends beyond academic achievement. She sees technology not just as a career but as a bridge to opportunity and dignity. Whether it is through wearable sensors, energy access, or future tools not yet imagined, her goal is to bring meaningful innovation to communities that have long been overlooked.

Her commitment is personal. Coming from a rural town in Colombia, Diana knows how transformative education and technology can be. That is why, whenever she visits home, she speaks at local schools, offering testimony that helps young students see what is possible. She also supports initiatives that help low-income children pursue higher education; small steps toward a bigger dream of one day building programs that integrate technology, equity, and education.

Diana imagines a future where children from villages like hers do not just benefit from technology—they shape it. Where curiosity is encouraged, creativity is nurtured, and access is never limited by geography or background. She hopes to be part of that change, not just through research, but by teaching, mentoring, and helping others believe in their potential.

Diana draws deep inspiration from the stories of her parents and grandparents, reflecting on the sacrifices they made to pursue education. Their strength fuels her own. In moments of doubt, when the weight of expectation feels heavy, she finds comfort in their unwavering belief in her. That love reminds her of why she started and where she wants to go.

Now, standing on the cusp of completion, surrounded by family, friends, and mentors who believe in her, Diana feels a profound sense of fulfillment. But this is only the beginning. With her heart set on service and her mind focused on impact, she steps forward, ready not only to shape her own future but also to inspire others to believe in theirs.

Breaking the Mold—My Personal Experience as a Woman in Engineering Technology

Colleen McGuiness

Introduction

I have an untraditional engineering technology experience. I graduated from Purdue University with a degree in Mechanical Engineering Technology, interned, and worked full-time in an engineering technology role, and then pivoted to law school to practice patent law. My background in engineering technology set me up for a successful career in patent law, as I can work with engineers and inventors to help protect their innovations. My experiences in engineering technology have also taught me important life lessons, which I hope others in the field can benefit from as well.

Background and Context

I was always drawn to the sciences as a kid. I joined every science club my schools offered, spent my days playing with Legos, and was constantly asking "How does this work?" I also loved playing with dolls, wearing tutus, and thought glitter went with everything. My brother, Jimmy, started college when I was 11 years old. I did not fully understand what he did at college, but I did know he attended Purdue University and studied engineering—a word I had only heard in passing up until this point. When he returned home for break, I found a stack of his textbooks in his bedroom, and as any little sister would, I started to look through them. I was fascinated by the equations and diagrams.

C. McGuiness (✉)
Purdue University, West Lafayette, IN, USA

A. Lucietto (ed.), *Innovating for Tomorrow: Engineering Technology*, Women in Engineering and Science, https://doi.org/10.1007/978-3-032-07645-8_16

Jimmy found me going through his textbooks and began to explain what engineering is and what a career in engineering looks like. Everything he said clicked with me, and I knew then I wanted to be an engineer. In the years that followed, I took every science and math elective my schools offered. During my junior year of high school, my school began to offer an engineering elective. I excitedly signed up but was shocked the first day when I saw that out of approximately 30 students, only 2 of us were girls. This class introduced me to the different types of engineering, taught me that I loved applied engineering, and the importance of women's representation.

I was initially attracted to engineering technology because of my love for science and math, as well as the enjoyment I felt in tackling difficult problems. Over time, this evolved to further include my desire to help global problems and as a creative outlet.

Personal Experiences and Stories

School came easily for me, at least it did until I started college. The professors covered the material quickly, the information did not click the first time, and I found myself with a significantly greater course load than I had previously experienced. I soon became overwhelmed and began to question if pursuing engineering was a mistake. I did enjoy the lectures and often found myself eager for the labs. But I felt like I did not belong because the material did not come as naturally to me as school had in the past.

Before starting at Purdue University, I was a straight-A student, tutored peers, and was able to balance a social life, sports, and work on top of my classes. My study habits from high school did not translate to college life, and after my first round of exams, it became clear to me that I was not excelling in the way I expected. I altered my study habits and tried to brute-force my way to success, but it proved to be futile. By the second month of school, it became apparent that I needed to seek help to navigate my engineering courses.

I met with my advisor, who provided me a list of tutors. I began to attend professor's office hours, met with TA's, and joined study groups. My grades began to improve, and the classes no longer felt overwhelming. I truly began to enjoy being challenged by my class work and viewed my confusion as an opportunity to learn and not a failure. This was the first, and arguably the greatest, lesson I learned in my pursuit of engineering technology—comfort does not breed growth, and not succeeding the first time is not failure but an opportunity to adapt and learn from those shortcomings. At first, I was anxious that others would view my need for extra help as a sign that I did not belong in the program. Once I received the help I needed, I realized that the most important thing was not what others thought of my abilities but that I received the assistance I needed for success.

Adjusting to the rigorous coursework was my first challenge in pursuing engineering technology, but it certainly was not my last. It became abundantly clear that

I did not look like many of my peers. Most of my engineering classes had only a handful of women, and in some of my classes, I was the only woman among my classmates. I tried not to focus on this; however, my immediate notice in a classroom was how many women sat among me. The most shocking realization was how many of my male peers were surprised that I was studying engineering technology.

During my sophomore year, I took a class where all the students were engineering technology majors. The disciplines of engineering technology varied, and because of this, I did not recognize all of my classmates. On the first day, I sat next to an unfamiliar face and introduced myself. He asked what my major was, and I replied with mechanical engineering technology. He paused, looking confused, and shared that despite this being a class for engineering technology majors, he did not think I was one. His reasoning was my appearance, specifically because I dressed "girly". I was shocked, as every student taking this class was required to be in the school of engineering technology. This was not the first, or last, time statements about my feminine style were made, but it was the most blatant. My peer did not hesitate to judge my appearance and deem that I could not be an engineering student and did not seem to find anything wrong in verbalizing it. Nor did he realize the impact of his words.

Throughout college and my career in engineering technology, I have often faced comments about my appearance and how that translates to my peers' perception of my capabilities. I considered trying to fit in more by dressing less feminine, not wearing makeup, and toning down my personality. I thought that if I put myself into the box my peers expected an engineer to fit, then maybe I would be more respected and questioned less. However, if I did change myself to stand out less, I would lose what made me who I am. Peers and even professors questioned my intelligence in a way I did not see them question my male classmates, but this did not hinder my success. There were times when these comments bothered me, but I did not let them stop me from wearing the clothes I wanted or dull my outgoing personality, and I am grateful I did not let others shrink me into a box.

This old-fashioned mindset followed me postgraduation when I began my career. I was the only woman in my position on the team. The team was careful to open meetings with "good morning, guys, and lady". The treatment between myself and my male counterparts was subtle but present. Leaders on my team were questioning team members about my relationship status and if I wanted to have children soon. Moreover, my technical knowledge was constantly questioned. Despite working with engines for my job, I would be asked if I knew how engines worked and how many years of experience I had.

I was, and still am, angered by the treatment that I and many other women in this industry experience. I am not alone in having my abilities questioned because of my appearance, my personality, or simply my gender. It is okay to be upset and frustrated because of this. Remember that you belong and never change yourself to fit the expectations of others.

It was around this time that I began to consider a career change. I was curious to see what other options I had. I spent a lot of time researching and thinking about what I want in a career. I loved engineering technology, but I also loved to read and write which were two areas lacking in my job. I heard about patent law and reached out to patent attorneys to schedule a meeting. I quickly realized that a career in patent law would let me blend my engineering technology background with my love for reading and writing. Additionally, this path would allow me to work with engineers and inventors and aid them in protecting their intellectual property rights—all of which resonated deeply with me.

Despite knowing that this transition was the right choice for me, it was difficult to leave engineering technology. Engineering was a goal I have had since middle school, and I had finally achieved that goal. I reflected on all the work I put in to get to this point and questioned if it was worth pivoting careers. Leaving my engineering job and starting law school was terrifying. I did not know if I would regret it and was intimidated by three more years of school. Ultimately, I am extremely happy that I made this change.

As I am preparing for graduation and ending my law school journey, I have realized that the lessons I learned in engineering technology have translated greatly. I no longer hesitate to ask questions or seek additional help. I have been my authentic self without fear of others' perceptions of me, and accepting change has opened an abundance of doors.

My journey to this point in my career and academics was not easy, nor did it look what I had expected it to. My advice to those who are pursuing a similar path: you belong here, do not be afraid to ask for help, and it is okay to change your mind. There will be a time when you walk into a classroom where no one else looks like you. There will be a time when a peer, professor, or coworker makes a biased comment about you. There are also so many people who are extremely happy to have you in this field and want you to succeed. It is okay to be angry or upset when someone tries to make you feel unwelcome, but remember what your purpose is and that you have already earned your place here.

You will struggle at one point or another in this industry, whether it is a difficult class, interview prep, or learning to balance school and life. Asking for help does not mean you are failing. Others will be able to guide you toward success and provide viewpoints that you may not have considered. You are not expected to know how to do everything, and this industry is filled with great support networks that are eager to help you.

The goals you set when you begin to pursue this path may change over time. This may look like a change in discipline or a complete career change. Trust yourself to make the right decision, and it is okay if that decision completely changes the path you are on.

Analysis and Discussion

Engineering technology provides a great base for your education. The material and lessons you learn not only prepare you to be a successful engineer but also transfer to all aspects of life. You will develop a strong work ethic, be able to think creatively, and possess great problem-solving skills. There are a multitude of career paths within the engineering technology field, but the skills you acquire in this industry translate to any other career path as well. An education and career in engineering technology are unlimited, and the skills you learn will transfer to all aspects of your life.

For me, my path took me toward patent law. An engineering technology degree prepared me for the rigorous academic environment of law school and future work as a patent attorney. I am able to combine my engineering background with my love for reading and writing. Engineering technology opens up so many doors and opportunities. Whether you pursue a career in engineering technology, engineering technology adjacent, or a completely different field, the skills you develop when pursuing an engineering technology degree will follow you for life.

Practical Insights and Recommendations

I recommend all students get as much experience as they can, do not be afraid to ask questions, and follow their instincts. Experience is vital for success. This is an opportunity to learn new things, hone your skills, and discover new interests. You are not expected to know everything—ask questions, join study groups, and attend office hours. You have an array of opportunities ahead of you. As long as you work hard and follow your passions, success will follow.

For professionals, always continue to learn, stand up for yourself, and follow your passion. Start each day with the goal of learning something new. There is always something to learn and improve on—be curious! Know that you do belong and that you are where you are because of your hard work. Do not let others make you question whether you deserve to be where you are. As your career develops, your interests, passions, and goals will change. Follow what calls to you and do not be afraid of change.

Professors, continue to adapt and support your students. Your students have different backgrounds, goals, and learning styles. What may work for one student may not for another. Focus on how to successfully reach each student and support their growth. Students will remember their professors for the rest of their lives. I will never forget the professors in my undergraduate and law school who have pushed me and invested in my success.

I have learned many lessons since I began my journey in engineering technology. The most important of which are to never be afraid to look foolish, follow your heart, and always give 100%. I wish I had asked questions during lectures, joined study groups, and attended office hours from the beginning. I was so focused on not wanting others to think I did not belong that I allowed myself to fall behind. Always seek assistance when needed. Asking for help does not mean you do not belong there—it will allow you to grow and succeed.

Always follow your heart. Your goals and dreams will change throughout your career, and do not be afraid to pursue them. I was scared to uproot my career and pivot into law. Now, I am incredibly grateful that I made this change and have a job that I am excited about. How you picture your career now is likely to change, which is okay. If you are scared to make a change, then do it afraid. Regardless of your academics or career, always give 100%. Be proud of your work and take pride in what you do. Take your opportunities seriously and give it your all. Do the study groups ask questions and take risks?

Conclusion

My experiences in engineering technology have truly shaped my life, and I am so grateful to have taken this path. Engineering technology was challenging, both academically and culturally. Peers have questioned whether I deserved to be in the position I was, and the classes challenged me in a way I had never experienced. I learned to ask for help, to not change who I was to fit in, and that it is okay to change your mind. I advise students to do their best but understand you are not supposed to know everything, try every experience that comes your way, and follow your passions. It is okay if your career or path looks different than others or different from what you expected. Keep working hard, and you will be successful.

Bridging Engineering and Community

Julia Miller Rowe

Educational Journey and Inspiration

Julia Rowe's passion for math and science began in high school, where she discovered her love for problem-solving. Her fascination with these subjects was not just academic; it was deeply rooted in her enjoyment of tackling complex issues and finding innovative solutions. Julia's ability to see a problem in its entirety and break it down into manageable parts was a skill that set her apart early on. This analytical mindset naturally led her to the field of engineering technology, where she could apply her problem-solving abilities in practical and impactful ways.

Julia found immense joy in the process of figuring out how to solve problems using various components and techniques. Her excitement for math and science was not limited to theoretical knowledge; she was eager to see how these subjects could be applied to real-world scenarios. This enthusiasm for practical application fueled her desire to pursue a career in engineering technology, where she could combine her love for problem-solving with her passion for innovation.

During her undergraduate studies, Julia faced both enjoyable and challenging moments. The rigorous coursework in engineering technology tested her resilience and determination. She encountered difficulties in keeping up with the demanding science courses and adapting to the traditional teaching methods. However, these challenges did not deter her; instead, they taught her valuable lessons about perseverance and self-discovery.

Julia's undergraduate experience was a journey of personal growth. She learned a lot about herself, including her strengths and areas for improvement. The

J. M. Rowe (✉)
Purdue University, West Lafayette, IN, USA

A. Lucietto (ed.), *Innovating for Tomorrow: Engineering Technology*, Women in Engineering and Science, https://doi.org/10.1007/978-3-032-07645-8_17

hands-on labs in her engineering technology program were particularly beneficial, as they allowed her to learn through practical experience. These labs helped Julia tangibly understand complex concepts, reinforcing her belief in the importance of hands-on learning.

Despite the difficulties, Julia's undergraduate experience shaped her future endeavors. She realized that her problem-solving knowledge could be used to make a meaningful impact in the world. This realization guided her career path and inspired her to explore how engineering principles could be applied to benefit communities. Julia's journey through undergraduate studies was a testament to her resilience, adaptability, and unwavering commitment to using her skills for the greater good.

Transition to Higher Education

After completing her undergraduate degree, Julia sought to apply engineering differently. She was particularly interested in how engineering is taught and wanted to make it more engaging for students. Julia struggled with the traditional methods of teaching engineering, which often emphasized rigorous structures and testing. She believed that students could thrive in engineering even if they were not top performers in math.

Julia's dissatisfaction with conventional teaching methods led her to explore alternative approaches to engineering education. She envisioned a more inclusive and engaging learning environment where students could excel regardless of their proficiency in math. Julia was passionate about demonstrating that engineering is not solely about excelling in math and science but that it is also about creativity, innovation, and problem-solving.

Her interest in community engagement further fueled her desire to transform engineering education. Julia began studying how students collaborate within service-learning courses, which integrate engineering principles with community service. These courses aim to apply engineering knowledge to real-world problems, benefiting both students and the community.

Julia's lifelong involvement in volunteer efforts, starting from her time in Girl Scouts, played a significant role in shaping her passion for community service. Her experiences in various volunteer organizations during college reinforced her belief in the power of community engagement. Julia saw firsthand how volunteer work could positively impact education and foster a sense of responsibility and empathy among students.

Her college experience solidified her interest in combining engineering with community service. Julia recognized that service-learning courses could provide students hands-on, practical experience while addressing community needs. This approach enhanced students' learning and instilled in them a sense of purpose and social responsibility.

Julia's journey into graduate school and completion of a Master of Science in Engineering Technology was driven by her desire to make engineering more accessible and impactful. She aimed to create an educational environment where students could thrive through practical application and community involvement. Julia's innovative approach to engineering education inspires her work and her commitment to using engineering for the greater good.

Overcoming Challenges

Significant challenges and triumphs marked Julia's academic journey. She struggled to keep pace with the demanding science courses during her undergraduate studies. The teaching methods employed in these courses were not always conducive to her learning style, which added to her struggles. One of the most significant hurdles she faced was severe test anxiety, which had a detrimental impact on her grades. Despite these obstacles, Julia discovered that hands-on labs in her engineering technology program were incredibly beneficial. These practical experiences allowed her to grasp complex concepts more effectively and boosted her confidence.

Graduate school brought a new set of challenges for Julia. Her health took a toll, necessitating a semester off to focus on recovery. This break, while essential, extended the duration of her master's program to 3 years. However, Julia's determination never wavered. She remained committed to her goals and worked diligently to complete her degree. Her support from the university staff, her friends, and the broader community was instrumental in her success. Their encouragement and assistance helped her navigate both her health issues and the academic demands she faced.

Ultimately, Julia's story is one of resilience and perseverance. Despite the numerous obstacles in her path, she managed to overcome them with the help of a supportive network and her own unwavering determination. Her journey highlights the importance of practical learning experiences, the impact of mental health on academic performance, and the value of a strong support system in achieving one's goals.

Mentors and Role Models

Her thesis advisor during graduate school profoundly influenced Julia's academic journey. This advisor played a crucial role in shaping her career path, offering academic guidance and mentorship that extended beyond the classroom. The advisor's support and encouragement were pivotal in helping Julia navigate the complexities of her graduate studies and in fostering her interest in engineering education.

A significant turning point in Julia's undergraduate studies occurred during a fluid mechanics class. Instead of assigning a traditional paper, her professor challenged her to design an entire course. This unconventional assignment was a revelation for Julia. It allowed her to explore the intricacies of teaching and curriculum development, sparking a passion for engineering education. She thrived in this environment, discovering a new dimension to her academic interests and career aspirations.

Throughout her graduate school journey, Julia was fortunate to have the support of several individuals who provided guidance and flexibility as she managed her medical issues. Their understanding and willingness to accommodate her needs were instrumental in her ability to persevere and succeed. Among these supporters were the workers at the campus utility plant, who played a particularly supportive role. They provided a strong foundation for her academic and personal growth, offering a stable and encouraging environment that helped her navigate her challenges.

Julia's story is a testament to the power of mentorship, innovative teaching methods, and the importance of a supportive community. Her experiences highlight how pivotal moments and influential individuals can shape one's career path and personal development. Despite the obstacles she encountered, Julia's resilience and the support she received enabled her to thrive and achieve her goals.

Current Role and Path Forward

Julia's current role is Director of Family Programs for the Witham YMCA. She is a "Social Architect," bridging the gap between high-level operational strategy and human-cnetered community impact. It is a perfect blend of her engineering background and her passion for community building. In this position, she oversees organizational strategic alliances, leverageing the innovative methods and design thinking principles she learned during her engineering studies to address the community's needs and impact. Her work involves a multifaceted approach to community engagement, where she gathers feedback from community members, designs initative, and oversees the creation of curricula tailored to community needs.

Julia's educational background in engineering, coupled with her master's degree, has equipped her with a robust set of skills that are invaluable in her current role. The problem-solving abilities she honed during her academic journey are now applied to real-world challenges, enabling her to devise creative solutions that foster community growth and development. Her expertise in design thinking allows her to approach community building with a strategic mindset, ensuring that the programs she develop are both innovative and impactful.

In her role, Julia finds immense joy in building better communities. She is passionate about making a positive difference and is dedicated to pursuing this path further. Her work benefits the community and provides her with a sense of fulfillment and purpose. Julia's journey is a testament to how interdisciplinary skills and a commitment to community service can lead to a rewarding and impactful career.

Julia exemplifies how technical expertise can be harnessed to create meaningful social change by continuously applying her engineering knowledge and design thinking to community engagement. Her story highlights the importance of adaptability, creativity, and a strong educational foundation to succeed in diverse fields.

Rewarding Projects

One of Julia's most rewarding projects has been building connections and creating community programs that directly benefit girls and their families. Her approach to community engagement is deeply rooted in understanding and addressing the community's specific needs. Julia has developed various programs that cater to these needs, ensuring they are both practical and impactful.

One such program involves providing bikes and transportation for girls, which has been a game-changer for many families. Julia recognized the importance of mobility and the freedom it brings, so she took the initiative to teach girls how to ride bikes. This empowers them with a valuable skill and promotes physical activity and independence. The joy and confidence of mastering bike riding have been transformative for many participants.

In addition to promoting physical activity, Julia has fostered excitement about reading and literature among the girls. She understands the power of literacy in opening doors to new opportunities and broadening horizons. By creating engaging reading programs, she has helped ignite a passion for books and learning, which has had a lasting positive impact on the community.

Julia's creativity shines through in her introduction of activities like using a Cricut machine. This innovative tool allows girls to explore their artistic capabilities and push their boundaries. Julia encourages them to think outside the box through these creative projects and develop new skills. The sense of accomplishment and pride that comes from creating something unique has been incredibly empowering for the girls.

Seeing the positive impact of these programs on the community has been incredibly fulfilling for Julia. Her dedication to making a difference and her ability to design programs that resonate with the community's needs have made her a beloved figure. Julia's work exemplifies how thoughtful, innovative approaches to community engagement can lead to meaningful and lasting change. Her journey is a testament to the power of empathy, creativity, and a strong commitment to serving others.

Professional Challenges

Julia's transition from corporate engineering to the nonprofit sector has presented unique challenges. Unlike the corporate world, where resources are often more abundant, the nonprofit space requires working with limited budgets and carefully prioritizing tasks to maximize impact. This shift has demanded a high level of resourcefulness and strategic thinking from Julia as she navigates the complexities of funding and resource allocation.

One of Julia's most significant obstacles is advocating for the community's needs and overcoming internal biases within the nonprofit organization. Advocacy in this context involves identifying the community's needs and effectively communicating these needs to stakeholders and decision-makers. Julia has had to become a vocal champion for the programs that will benefit the community, often pushing for initiatives that may not initially receive widespread support.

Overcoming internal biases has been a significant part of her role. These biases can manifest in various forms, such as preconceived notions about what programs are most valuable or skepticism about the feasibility of specific initiatives. Julia's ability to challenge these biases and advocate for the community's needs has driven positive change. Her persistence and dedication to ensuring that the community's voice is heard have been instrumental in securing support for vital programs.

Despite these challenges, Julia's passion for community engagement and commitment to making a difference have motivated her. She has developed a strong sense of resilience and adaptability, learning to navigate the nonprofit landscape with determination and creativity. Her efforts to advocate for the community and implement impactful programs have benefited those she serves and strengthened her professional skills and resolve.

Julia's journey in the nonprofit sector highlights the importance of advocacy, resourcefulness, and overcoming internal biases. Her experiences underscore the challenges and rewards of working in a space where the primary focus is making a meaningful difference in people's lives. Through her unwavering dedication, Julia continues to push boundaries and create positive change, proving that impactful work is possible despite significant obstacles.

Unique Perspective as a Woman in Engineering Technology

Julia's unique perspective in engineering technology is characterized by her ability to blend technical expertise with a strong focus on community innovation. Unlike many professionals who may prioritize technological advancements for their own sake, Julia constantly considers how her engineering skills can benefit her community. This community-oriented approach significantly enhances her work's value, bridging the gap between technology and social impact.

Julia's approach emphasizes the importance of understanding the broader implications of engineering projects on consumers and the community. She recognizes that engineering is not just about creating new technologies but also about ensuring these innovations positively and meaningfully impact people's lives. By fostering excitement around engineering projects and highlighting their benefits, Julia engages in the community and encourages a collaborative spirit.

Her mindset allows her to use her innovation and expertise to make a tangible difference. For example, when developing new programs or initiatives, Julia considers how these projects can address specific community needs, improve quality of life, and promote sustainable development. This holistic view ensures that her engineering solutions are technically sound and socially responsible.

Julia's work exemplifies the potential of engineering to drive social change. By prioritizing community needs and focusing on the human aspect of technology, she creates projects that resonate with people and address real-world challenges. Her ability to inspire others and build excitement around engineering further amplifies the impact of her work, as it encourages more individuals to get involved and contribute to community-driven innovation.

In summary, Julia's approach to engineering technology is a powerful example of how technical skills can be harnessed to create social value. Her dedication to understanding and addressing community needs, combined with her innovative mindset, allows her to make a positive impact beyond traditional technological advancements. Through her work, Julia demonstrates the importance of integrating engineering with social responsibility to build a more inclusive future.

Gender-Related Challenges

Julia's journey through her undergraduate studies and internships was marked by significant gender-related challenges. Throughout her academic and professional experiences, she often faced assumptions about her capabilities simply because of her gender. These assumptions meant that Julia had to repeatedly prove her knowledge and skills, a task that required considerable resilience and determination.

Her practical experience began early, working in a machine shop during high school. This hands-on experience was invaluable, providing her with a solid foundation in engineering and technical skills. However, even in this environment, Julia encountered over-explanations from male peers who underestimated her abilities. These interactions were frustrating and highlighted the pervasive gender biases that exist in technical fields.

As a young professional under 30, Julia has continued to face challenges in being taken seriously, especially when working alongside older colleagues. Age-related biases can compound gender-related challenges, making it even more difficult for young women to assert their expertise and gain the respect they deserve. Julia's experiences underscore the importance of addressing these biases and creating more inclusive environments where all individuals are valued for their contributions.

Despite these obstacles, Julia's perseverance and commitment to her field have enabled her to overcome many challenges. She has demonstrated her capabilities time and again, earning the respect of her peers and colleagues through her hard work and dedication. Julia's story is a powerful reminder of the need for continued efforts to combat gender and age biases in STEM fields, ensuring that talented individuals like her can thrive and make meaningful contributions without undue barriers.

Her experiences also highlight the importance of mentorship and support networks in helping individuals navigate these challenges. By fostering a culture of inclusivity and respect, organizations can help mitigate the impact of biases and create environments where everyone can succeed. Julia's journey inspires others facing similar challenges, showing that resilience and determination can lead to success even in the face of adversity.

Observations on Gender Diversity

Julia's career has been marked by notable changes in gender diversity within the engineering industry. During her internships, she often was the only female in the engineering department. This lack of gender diversity sometimes led to uncomfortable situations, where Julia had to repeatedly navigate a predominantly male environment and prove her capabilities. These experiences highlighted women's challenges in male-dominated fields and underscored the need for greater inclusivity.

After graduation, Julia's work with a girl coalition starkly contrasted with her previous experiences. This supportive environment, characterized by female leadership, offered Julia a sense of belonging and validation she had not always felt during her internships. The presence of strong female role models and mentors in the coalition helped Julia thrive, as she no longer had to fight as hard to be taken seriously. This empowering shift allowed her to focus more on her work and less on overcoming gender biases.

However, Julia has noticed that age-related challenges have become more prominent in her current role. As a young professional under 30, she sometimes has to assert her ideas more strongly to be heard, especially when surrounded by older colleagues. This dynamic requires Julia to push harder to establish herself as an equal among her peers and to ensure her contributions are recognized. The need to assert her ideas and navigate power struggles has been a significant aspect of her professional journey.

The transition from a male-dominated environment during her internships to a female-led workplace in her professional career has been interesting for Julia. While the supportive environment has alleviated some gender-related challenges, it has introduced new dynamics that require her to prove her worth and establish her authority continuously. Julia's ability to adapt to these changing environments and assert her ideas demonstrates her resilience and determination.

Julia's experiences highlight the evolving landscape of gender diversity in engineering and the importance of supportive environments in fostering professional growth. Her journey underscores the need for continued efforts to address both gender and age biases, ensuring that all individuals have the opportunity to succeed and be heard. Through her perseverance and adaptability, Julia continues to make a positive impact in her field, advocating for inclusivity and recognition for all.

Advice for Young Women in Engineering Technology

Julia's advice to young women considering a career in engineering technology is both empowering and insightful. She encourages them to broaden their understanding of engineering, emphasizing that it is more than just math and science. Engineering is a versatile field that applies to many different areas, from healthcare to environmental sustainability, and the skills learned in engineering can be incredibly valuable in various contexts.

One of Julia's key points is that engineering is fundamentally about innovation, design, and problem-solving. While math and science are important components, they are not the sole determinants of success in engineering. Julia reassures young women that they do not need to be perfect at these subjects to pursue a career in engineering. Instead, they should focus on developing their creativity, critical thinking, and ability to tackle complex problems.

Julia highlights the importance of design thinking, a methodology that involves understanding users' needs, brainstorming solutions, prototyping, and testing. These skills are applicable not only in engineering but in many aspects of life and work. By mastering design thinking, young women can approach challenges strategically and create innovative solutions that make a difference.

Moreover, Julia advises young women to explore the diverse fields within engineering technology. There are numerous paths to choose from, whether it is biomedical engineering, civil engineering, or software engineering. Each field offers unique opportunities to apply engineering principles in ways that can profoundly impact society.

Julia's message is clear: engineering is a dynamic and rewarding career that welcomes creativity and innovation. She encourages young women to embrace their strengths, pursue their interests, and not be deterred by the misconception that engineering is solely about math and science. By focusing on problem-solving and design, they can carve out successful and fulfilling careers in engineering technology.

In summary, Julia advises young women to view engineering as a field rich with possibilities and approach it with confidence and curiosity. Her emphasis on innovation and design thinking provides a refreshing perspective that can inspire the next generation of female engineers to make their mark on the world.

Staying Current in Engineering Technology

Although Julia has shifted her focus toward community engagement, she remains deeply involved in STEM education. She understands the importance of staying current with technological advancements, particularly in the rapidly evolving field of engineering technology. Julia closely monitors how artificial intelligence (AI) impacts teachers and higher education in the engineering space. AI has the potential to revolutionize education by personalizing teaching methods, providing prompt feedback, and automating administrative tasks. Julia is particularly interested in how educators and students are adapting these technologies, as they can significantly enhance the learning experience and improve educational outcomes.

Julia's interest in AI extends to its applications in teaching and learning. She observes how AI-driven tools, such as adaptive learning platforms and automated grading systems, are being integrated into engineering education. These technologies offer personalized learning experiences, allowing students to engage with materials more effectively. For educators, AI simplifies complex concepts and provides real-time feedback, enabling them to focus on higher-level teaching and mentoring.

In addition to AI, Julia keeps up with other emerging technologies in engineering education, such as virtual reality (VR) and augmented reality (AR). These technologies create immersive learning environments where students can experiment with engineering principles without physical space limitations. VR and AR offer hands-on experiences crucial for understanding complex concepts and fostering student collaboration.

Julia's commitment to being informed about these developments reflects her dedication to her community engagement work and her passion for STEM education. By keeping up with the latest advancements, she ensures that her programs and initiatives are informed by cutting-edge technologies, ultimately benefiting the community she serves. Her proactive approach to learning and adapting new technologies exemplifies her innovative mindset and her desire to make a positive impact through her work.

Julia's active continuing involvement in STEM education and her keen interest in AI and other emerging technologies highlight her commitment to staying current in engineering technology. Her efforts to understand and integrate these advancements into her work demonstrate her dedication to enhancing educational experiences and fostering community growth.

Essential Skills for Success

Julia believes that creativity is a cornerstone for future success in engineering technology. She challenges the common assumption that proficiency in math and science is the sole determinant of success in this field. While these subjects are essential, Julia emphasizes that they are not the only skills that matter. Instead, she highlights the significance of creative problem-solving, big-picture thinking, and the ability to break down complex problems into manageable steps.

Creative problem-solving involves approaching challenges with an open mind and thinking outside the box to find innovative solutions. Julia believes that this skill is crucial for engineers, as it allows them to develop unique and effective solutions to real-world problems. By fostering creativity, engineers can push the boundaries of what is possible and drive technological advancements.

Big-picture thinking is another essential skill that Julia advocates for. This involves understanding the broader context of a problem and considering how different elements interact with one another. Engineers who can see the big picture are better equipped to design systems and solutions that are not only effective but also sustainable and scalable. Julia believes that this holistic approach is vital for addressing complex challenges and making a lasting impact.

The ability to break down problems into logical steps is also a key skill for engineers. Julia emphasizes that engineering is about more than just solving equations; it involves a systematic approach to problem-solving. By breaking down complex problems into smaller, more manageable parts, engineers can tackle each component methodically and ensure that their solutions are robust and reliable.

Julia's thoughts on future success in engineering technology revolve around the importance of creativity, big-picture thinking, and systematic problem-solving. She encourages aspiring engineers to cultivate these skills and recognize that engineering success is not solely dependent on math and science proficiency. By embracing creativity and a holistic approach to problem-solving, engineers can make meaningful contributions to their field and drive innovation forward.

Future Aspirations

Julia's future aspirations are deeply rooted in her passion for nonprofit work and education. She envisions continuing to advance in these fields, driven by her commitment to positively impacting young people's lives. One of her key goals is to further her education, focusing on researching how problem-solving skills can be effectively taught to younger students. Julia believes that instilling these skills early on can profoundly affect children's development and future success.

Her interest in problem-solving extends beyond traditional academic subjects. Julia is passionate about developing innovative programs that engage youth in learning through hands-on, practical experiences. She aims to create educational initiatives that teach problem-solving and inspire creativity and critical thinking. By designing educational and enjoyable programs, Julia hopes to foster a lifelong love of learning in young students.

Julia's vision includes collaborating with educators, researchers, and community leaders to develop and implement these programs. She understands the importance of a multidisciplinary approach and values the insights and expertise that others can bring to her initiatives. Through these collaborations, Julia aims to create a supportive and enriching environment for students, where they can explore their interests and develop essential skills.

In addition to her focus on youth education, Julia is also committed to continuing her work in community engagement. She aspires to build stronger connections between educational institutions and the communities they serve, ensuring that educational programs are responsive to the community's needs and aspirations. By bridging the gap between education and community development, Julia hopes to create a more inclusive and supportive environment for all.

Julia's future aspirations reflect her dedication to education, innovation, and community service. She aims to make a lasting impact by empowering young people with the skills and confidence they need to succeed and to inspire the next generation of problem-solvers and innovators through her continued efforts in the nonprofit and education sectors.

Evolving Role of Women in Engineering Technology

Julia envisions a future where women are actively involved in every technology sector, from computer science to structural engineering. She believes that the unique perspectives women bring to the table are invaluable. Having faced and overcome various challenges, women possess insights that can drive innovation and foster inclusive growth in the tech industry.

Julia is passionate about seeing women take over the tech space and be present in all areas of engineering and technology. She recognizes that diversity in thought and experience leads to more robust and creative solutions. Women, with their diverse backgrounds and experiences, can contribute significantly to advancing technology and engineering.

Julia hopes to see more women in leadership roles within the tech industry, where they can influence decision-making and shape the future of technology. She believes that women have the potential to lead groundbreaking projects, mentor the next generation of engineers, and create a more equitable and inclusive environment.

Her vision includes initiatives supporting and encouraging women to pursue technology and engineering careers. Julia advocates for mentorship programs, scholarships, and networking opportunities to help women navigate their professional journeys and achieve their goals. She also emphasizes the importance of creating supportive workplaces that value diversity and provide equal opportunities for all.

Julia's aspirations for women in technology are rooted in her belief in the power of diversity and inclusion. She envisions a tech industry where women are present and thriving, making significant contributions and leading toward a more innovative and equitable future. Julia hopes to inspire and empower women to take their rightful place in engineering and technology through her advocacy and efforts.

Supporting Women in Engineering

Julia's commitment to advancing women's rights and promoting gender equality is a central theme in her life. She remains actively involved with the Society of Women Engineers (SWE), an organization dedicated to empowering women in engineering and technology. Through her participation in SWE, Julia contributes to initiatives that support women in STEM fields, providing mentorship, networking opportunities, and advocacy for gender equity.

Beyond her involvement with SWE, Julia is passionate about women's rights. She participates in various organizations focusing on women's rights in the workplace and beyond. These organizations work to address issues such as gender discrimination, workplace harassment, and pay equity, striving to create safer and more inclusive environments for everyone. Julia's dedication to these causes reflects her belief in creating equitable opportunities and fostering a culture of respect and inclusion.

Julia's efforts to make communities safer for everyone are driven by her desire to see positive change and her commitment to social justice. She understands that achieving gender equality requires collective action and is determined to contribute to this movement. By advocating for women's rights and participating in organizations that champion these causes, Julia is vital in advancing the conversation around gender equity and creating a more just society.

In conclusion, Julia's background includes resilience, innovation, and advocacy. Significant challenges and triumphs have marked her journey through engineering technology and community engagement. Julia's dedication to empowering women, promoting gender equality, and making communities safer is a testament to her unwavering commitment to social justice. As she continues to advance in her career and pursue her passions, Julia's impact on the world around her will undoubtedly grow, inspiring others to follow in her footsteps and contribute to a more equitable and inclusive future.

Navigating the Path: Insights from a Systems Engineer

Meher Taleyarkhan

Introduction

Engineering technology is a dynamic and multifaceted field crucial to advancing modern industry. It bridges the gap between theoretical engineering principles and practical applications, making it indispensable in developing and implementing innovative solutions. Despite its significance, the field has historically been male-dominated, presenting unique challenges and opportunities for women pursuing careers in engineering technology.

This chapter delves into Meher Taleyarkhan's journey as a systems engineer whose experiences provide valuable insights into the educational pathways, professional challenges, and strategies for success in engineering technology. Through her narrative, we explore the pivotal moments that shaped her career, the obstacles she overcame, and her contributions to her field.

M. Taleyarkhan (✉)
Purdue University, West Lafayette, IN, USA

A. Lucietto (ed.), *Innovating for Tomorrow: Engineering Technology*, Women in Engineering and Science, https://doi.org/10.1007/978-3-032-07645-8_18

Meher's story begins with a childhood fascination with roller coasters, which ignited her passion for engineering. This early interest set her on a path of academic and professional exploration, leading her to pursue a degree in mechanical engineering technology at Purdue University. Her journey was not without its hurdles; she faced academic challenges, struggled with maintaining a high GPA, and encountered gender-related obstacles in the workplace. However, her resilience and determination enabled her to navigate these challenges and achieve success in her career.

In her role as a systems engineer at Northrop Grumman, Meher integrates various engineering disciplines to ensure cohesive project outcomes. Her ability to bridge gaps between different teams and facilitate effective communication underscores the critical role of systems engineering in advancing technology. Moreover, her experiences highlight women's unique insights and perspectives in the field, enhancing team dynamics and fostering a more inclusive work environment.

This chapter also addresses the broader context of gender diversity in engineering technology. While there has been progress in increasing the representation of women in technical roles, significant disparities remain, particularly in leadership positions. Meher's observations and experiences provide a lens through which we can examine these issues and consider strategies for fostering greater gender diversity and inclusion in the field.

As we explore Meher's journey, we will also discuss the importance of work–life balance, the value of soft skills, and the need for continuous learning and adaptation in a rapidly evolving technological landscape. Her story serves as both an inspiration and a guide for aspiring women engineers, offering practical advice and insights for navigating the path to success in engineering technology.

Educational Journey

Meher's interest in engineering was sparked during middle school, when curiosity and exploration often shape future aspirations. Her fascination with roller coasters and their mechanisms for ensuring safety and fun played a pivotal role in igniting her passion for engineering. This early interest was nurtured through educational experiences emphasizing hands-on learning and real-world applications.

Early Inspiration

Meher's journey began with family trips to theme parks in Florida, notably Universal Studios, where she encountered thrilling roller coasters. Initially, these rides instilled a sense of fear, but a middle school science class on potential and kinetic energy transformed her perspective. The class included roller coaster simulations, which helped Meher visualize the principles of energy in action. This newfound understanding alleviated her fear and sparked a deep fascination with the engineering behind roller coasters and a pursuit to one day design thrill rides to provide the same joyful experience to others.

High School Years

During her high school years, Meher's interest in engineering grew. She excelled in science and mathematics, subjects that are foundational to engineering. Her high school offered various STEM (Science, Technology, Engineering, and Mathematics) programs and extracurricular activities, which allowed her to explore different engineering aspects. Participation in science fairs and engineering workshops further solidified her desire to pursue a career in this field.

Choosing a College and Major

When it came time to apply for college, Meher was determined to pursue mechanical engineering, which offered many opportunities. She knew of Purdue University, which is known for its strong engineering programs. However, gaining admission to the School of Engineering proved challenging, and Meher faced initial setbacks. Despite these obstacles, she was accepted into the School of Engineering Technology, a major she was initially unfamiliar with.

Discovering Engineering Technology

Engineering technology was a perfect fit for Meher's learning style and interests. Unlike traditional engineering programs, which focus on theoretical concepts, engineering technology emphasizes practical applications and hands-on learning. Meher thrived in this environment, where she could directly apply what she learned in the classroom to real-world scenarios. Courses in metallurgy, materials testing, and mechanical systems gave her a comprehensive understanding of applied engineering principles.

Undergraduate Research

Recognizing the importance of gaining practical experience, Meher sought opportunities for undergraduate research. Her GPA was not initially competitive for internships, so she turned to research as an alternative. She became involved in various research projects related to engineering education and renewable energy systems. These experiences enhanced her technical skills and sparked an interest in research and theoretical analysis.

Overcoming Academic Challenges

Meher's academic journey was marked by perseverance and resilience. She faced challenges in maintaining a high GPA and struggled with the rigorous demands and newness of her coursework. However, she found support from her professors and peers, who encouraged her to keep pushing forward. Through hard work and determination, Meher improved her academic performance and gained confidence in her abilities.

Graduate Studies

After completing her undergraduate degree, Meher considered pursuing further education to enhance her career prospects. Encouraged by her undergraduate professor, who highlighted the benefits of advanced studies, Meher applied for a master's program in engineering technology at Purdue University and was accepted. During her graduate studies, her master's chair/advisor (the same professor from her undergraduate years) further motivated her, emphasizing the importance of a technical master's degree increased her odds of securing a job. Graduate studies allowed her to delve deeper into her areas of interest, particularly the intersection of engineering and business. Her thesis project involved a financial analysis of renewable energy systems at a wastewater plant, combining her technical expertise with her growing interest in business and management.

Transition to Professional Life

Upon graduating with her master's degree, Meher faced a critical decision: whether to continue her education with a PhD or enter the workforce. She attended the Society of Women Engineers (SWE) Conference and interviewed and received a job offer from Northrop Grumman as a systems engineer. This role allowed her to apply her technical skills in a practical setting. This transition marked the beginning of her professional journey, where she would continue to learn, grow, and contribute to engineering technology.

Initial Setbacks

Meher's first major challenge came when she applied to Purdue University's School of Engineering. Despite her passion for engineering and strong academic background, she faced rejection in her initial attempts to gain admission. This setback significantly blew her confidence, as she had set her sights on a traditional engineering program. However, instead of giving up, Meher explored alternative options and discovered the School of Engineering Technology.

Adjusting to a New Major

Transitioning to engineering technology required Meher to adjust her expectations and embrace a different approach to learning. Engineering technology focuses on practical applications and hands-on experiences, perfectly fitting Meher's learning style. She thrived in an environment where she could directly apply theoretical concepts to real-world scenarios. This shift reignited her passion for engineering and gave her a unique perspective on the field.

Struggling with GPA

Maintaining a high GPA was another significant challenge for Meher. The rigorous demands of her coursework and the pressure to perform well academically often left her feeling overwhelmed. Despite her best efforts, her GPA did not initially meet the competitive standards required for internships and other opportunities. This was a critical moment in her academic journey, as she had to find ways to improve her performance and demonstrate her capabilities.

Seeking Support and Mentorship

Recognizing the need for support, Meher contacted her professors and peers for guidance. She found mentors who provided valuable advice and encouragement, helping her navigate the complexities of her coursework. One of her key mentors was Professor Lucietto, who offered career advice and helped Meher identify growth opportunities. This support network played a crucial role in boosting her confidence and academic performance.

Engaging in Undergraduate Research

With her GPA not initially competitive for internships, Meher turned to undergraduate research as an alternative way to gain practical experience. She became involved in various research projects related to engineering education and renewable energy systems. These experiences enhanced her technical skills and sparked an interest in research and theoretical analysis. Engaging in research allowed Meher to apply her knowledge meaningfully and build a strong foundation for her future career.

Improving Academic Performance

Through hard work and determination, Meher gradually improved her academic performance. She adopted effective study strategies, such as time management, active learning techniques, and seeking help when needed. Her persistence paid off, and she began to see positive changes in her GPA and overall academic standing. This improvement opened up new opportunities, including the chance to pursue graduate studies.

Graduate Studies and Thesis Project

After completing her undergraduate degree, Meher decided to further her education by applying for a master's program in engineering technology at Purdue University. She was accepted into the program, which allowed her to delve deeper into her areas of interest. Her thesis project involved a financial analysis of renewable energy systems at a wastewater plant, combining her technical expertise with her growing interest in business and management. This project allowed Meher to explore the intersection of engineering and business, further expanding her skill set.

Transition to Professional Life

Upon graduating with her master's degree, Meher faced a critical decision: whether to continue her education with a PhD or enter the workforce. She chose to accept a job offer from Northrop Grumman as a systems engineer, a role that allowed her to apply her technical skills in a practical setting. This transition marked the beginning of her professional journey, where she would continue to learn, grow, and contribute to the field of engineering technology.

Professional Role and Contributions

Meher's professional journey as a systems engineer at Northrop Grumman is a testament to her ability to integrate various engineering disciplines and ensure cohesive project outcomes. Her role involves a blend of technical expertise, leadership, and effective communication, making her an invaluable asset to her team and the broader organization.

As a systems engineer, Meher's primary responsibility is to act as a bridge between different engineering disciplines, ensuring that all components of a project work harmoniously. Systems engineering is a multidisciplinary field that requires a broad understanding of mechanical, electrical, and software engineering principles. Meher's role involves coordinating these diverse elements to create a unified and functional system.

In her position, Meher works closely with mechanical, electrical, and software engineers, ensuring that their individual contributions align with the overall project goals. This involves understanding the specific requirements and constraints of each discipline and facilitating effective communication between teams. Her ability to manage complex projects and keep teams on track is crucial to the success of the projects she oversees.

One of the critical aspects of systems engineering is identifying gaps and potential issues before they become significant problems. Meher excels at this, using her broad knowledge base to spot inconsistencies and areas where different components may not align. She then works with the relevant teams to address these gaps, ensuring a smooth integration process.

Meher is also involved in implementing model-based systems engineering (MBSE) practices, which use digital models to represent and analyze system components and their interactions. This approach helps in visualizing complex systems, identifying potential issues early, and improving overall project efficiency. Meher's expertise in MBSE tools, such as Cameo from CATIA, allows her to create detailed models that facilitate better decision-making and communication among teams.

Effective communication with customers and stakeholders is a vital part of Meher's role. She ensures that project requirements are clearly understood and met, and she provides regular updates on progress. Her ability to articulate technical details in a way that is accessible to non-engineers is a key strength, helping to build trust and ensure alignment with customer expectations.

Meher's contributions to the field are multifaceted. Her work on integrating renewable energy systems at a wastewater plant is an example of her innovative approach to engineering challenges. By combining her technical expertise with financial analysis, she contributed to a project that not only advanced engineering practices but also promoted sustainability.

Beyond her technical contributions, Meher plays a significant role in mentoring junior engineers and fostering a collaborative work environment. She provides guidance and support to new team members, helping them navigate the complexities of their roles and develop their skills. Her leadership style emphasizes inclusivity and teamwork, creating a positive and productive atmosphere.

As a woman in a predominantly male field, Meher serves as a role model and advocate for gender diversity in engineering. She actively participates in initiatives aimed at supporting and empowering women in engineering, sharing her experiences and insights to inspire the next generation of female engineers.

Meher is committed to continuous improvement, both in her professional development and in the projects she works on. She stays abreast of the latest trends and technologies in engineering, ensuring that her skills remain relevant and that she can bring the best practices to her work. Her dedication to learning and growth is reflected in the quality and impact of her contributions.

One of Meher's notable projects involved conducting a financial and engineering analysis of renewable energy systems at a wastewater plant. This project required a deep understanding of both technical and financial aspects, showcasing her ability to integrate diverse knowledge areas to achieve sustainable solutions.

Meher has been instrumental in pushing the adoption of digital architectures and MBSE practices within her team. By creating detailed digital models, she has improved the efficiency and accuracy of system integration processes, leading to better project outcomes and reduced risks.

In her role as a deputy systems engineering integration and test (SEIT) lead, Meher works closely with chief engineers and program managers to ensure that projects meet their technical and compliance requirements. Her leadership in this capacity has been critical in maintaining project alignment and achieving successful deliverables.

Unique Insights and Perspectives

Women in engineering technology bring unique insights and perspectives that significantly enhance team dynamics, project outcomes, and the overall work environment. These contributions are often rooted in their ability to understand and navigate complex interpersonal relationships, foster inclusive collaboration, and approach problem-solving with empathy and creativity.

Emotional Intelligence and Team Dynamics

One of the key strengths that women often bring to engineering technology is their heightened emotional intelligence. Emotional intelligence involves the ability to recognize, understand, and manage one's own emotions, as well as the emotions of others. This skill is particularly valuable in team settings, where effective communication and collaboration are essential.

Meher notes that women are typically more cognizant of emotions and better at reading the room. This ability allows them to gauge the mood and dynamics of a group, ensuring that all team members feel heard and valued. In engineering projects, where diverse perspectives and expertise are crucial, fostering an inclusive environment can lead to more innovative solutions and better decision-making.

Inclusive Collaboration

Women in engineering technology often excel at creating inclusive and collaborative work environments. Meher observes that women engineers are more likely to engage with quieter team members, actively seeking their input and ensuring that their voices are heard. This approach not only enhances team cohesion but also leverages the full range of skills and knowledge within the group.

Inclusive collaboration is particularly important in multidisciplinary projects, where the integration of different engineering disciplines is required. By encouraging open communication and valuing diverse perspectives, women engineers help to break down silos and facilitate more effective teamwork. This leads to more comprehensive and well-rounded solutions to complex engineering challenges.

Effective Communication

Effective communication is another area where women in engineering technology often excel. Meher highlights that women engineers tend to be more adept at conveying problematic updates in a less blunt and more constructive manner. This skill is crucial in maintaining positive working relationships and ensuring that issues are addressed without causing unnecessary conflict.

In her role as a systems engineer, Meher's ability to articulate technical details in a way that is accessible to non-engineers is a key strength. This skill helps to build trust with customers and stakeholders, ensuring that project requirements are clearly understood and met. Effective communication also facilitates better collaboration among different engineering teams, leading to more cohesive and successful project outcomes.

Empathy and Problem-Solving

Empathy is a powerful tool in problem-solving, allowing engineers to consider the human impact of their work and approach challenges with a broader perspective. Women in engineering technology often bring a strong sense of empathy to their roles, which can lead to more thoughtful and user-centered solutions.

Meher's empathetic approach is evident in her work on integrating renewable energy systems at a wastewater plant. By considering both the technical and financial aspects of the project, she was able to develop solutions that not only advanced engineering practices but also promoted sustainability and community well-being.

Leadership and Mentorship

Women in engineering technology also play a crucial role in leadership and mentorship. Meher's leadership style emphasizes inclusivity and teamwork, creating a positive and productive atmosphere. She provides guidance and support to junior engineers, helping them navigate the complexities of their roles and develop their skills.

Mentorship is particularly important in fostering the next generation of engineers. By sharing their experiences and insights, women leaders like Meher inspire and empower young engineers to pursue their passions and overcome challenges. This mentorship helps to build a more diverse and inclusive engineering community, where all voices are valued and respected.

Advocacy for Gender Diversity

As a woman in a predominantly male field, Meher serves as a role model and advocate for gender diversity in engineering. She actively participates in initiatives aimed at supporting and empowering women in engineering, sharing her experiences and insights to inspire the next generation of female engineers.

Continuous Improvement and Learning

Women in engineering technology are often committed to continuous improvement and lifelong learning. Meher exemplifies this commitment by staying abreast of the latest trends and technologies in engineering. She leverages social media platforms like LinkedIn for professional updates and participates in internal technical career fairs and lunch-and-learn sessions.

This dedication to learning ensures that women engineers remain at the forefront of their field, bringing the best practices and innovative solutions to their work. Continuous improvement also fosters a culture of excellence, where engineers are encouraged to push boundaries and strive for the highest standards in their projects.

Gender-Related Challenges

Despite women's significant contributions to engineering technology, they often face gender-related challenges that can impact their career progression and overall experience in the field. These challenges include biases, stereotypes, and systemic barriers that require resilience and strategic approaches to overcome.

Bias and Stereotyping

One of women's most pervasive challenges in engineering technology is bias and stereotyping. These biases can manifest in various ways, from subtle microaggressions to overt discrimination. Women may be perceived as less competent or less committed to their careers compared to their male counterparts, based on gender stereotypes.

Meher's experiences reveal the difficulties of being steamrolled in competitive environments, where louder and more assertive voices often dominate. This can be particularly challenging for women who are more introverted or soft-spoken, as their ideas and contributions may be overlooked or undervalued. Overcoming these biases requires women to assert themselves confidently and demonstrate their expertise and value.

Navigating Competitive Environments

Engineering technology is a highly competitive field, and women often have to navigate environments where assertiveness and self-promotion are key to success. Meher's journey highlights the need to speak out in meetings and group settings, ensuring that her ideas are heard and respected. This can be a daunting task, especially for those who are naturally reserved.

Strategies such as preparing thoroughly for meetings, practicing assertive communication, and seeking allies within the team can help women navigate these competitive environments. Building a network of supportive colleagues and mentors can also provide valuable guidance and encouragement.

Balancing Assertiveness and Collaboration

Women in engineering technology often face the challenge of balancing assertiveness with collaboration. While it is important to assert oneself and advocate for one's ideas, it is equally important to foster a collaborative and inclusive work environment. Meher's approach involves speaking out confidently and respectfully, ensuring that her contributions are recognized without alienating her colleagues.

Effective communication skills are crucial in achieving this balance. Women can benefit from training in assertive communication techniques, which help them articulate their ideas clearly and confidently while maintaining positive relationships with their peers. Additionally, seeking feedback and continuously improving communication skills can enhance their ability to navigate complex team dynamics.

Overcoming Systemic Barriers

Systemic barriers, such as unequal access to opportunities and resources, can also hinder women's progress in engineering technology. These barriers may include limited access to mentorship, fewer networking opportunities, and biases in hiring and promotion processes. Meher's experiences underscore the importance of resilience and adaptability in overcoming these obstacles.

Women can address systemic barriers by actively seeking out mentorship and networking opportunities, advocating for themselves in performance reviews and promotions, and participating in professional organizations that support gender diversity. Engaging in initiatives aimed at promoting diversity and inclusion within their organizations can also help create a more equitable work environment.

Gender Diversity in Leadership

While there has been progress in increasing the representation of women in technical roles, significant disparities remain, particularly in leadership positions. Meher observes that women tend to gravitate toward managerial roles, but there is a growing need for female representation in technical leadership positions, such as chief engineers.

Promoting gender diversity in leadership requires organizations to implement policies and practices that support women's career advancement. This includes providing leadership training, creating mentorship programs, and ensuring that promotion processes are fair and transparent. Women can also take proactive steps to develop their leadership skills and seek out opportunities for advancement.

Advocacy and Support Networks

Advocacy and support networks play a crucial role in addressing gender-related challenges. Meher actively participates in initiatives aimed at supporting and empowering women in engineering, sharing her experiences and insights to inspire the next generation of female engineers. These efforts contribute to raising awareness about the importance of diversity and inclusion and create opportunities for women to thrive in engineering technology.

Professional organizations, conferences, and panels focused on gender diversity provide valuable platforms for women to connect, share experiences, and advocate for change. By engaging with these networks, women can access resources, mentorship, and support that help them navigate their careers and overcome challenges.

Personal Strategies for Success

Meher's personal strategies for success include speaking out confidently, seeking mentorship, and continuously improving her skills. She emphasizes the importance of resilience and adaptability, recognizing that setbacks are a natural part of the journey. By maintaining a positive mindset and focusing on her goals, Meher has been able to overcome gender-related challenges and achieve success in her career.

Women in engineering technology can benefit from adopting similar strategies, including setting clear career goals, seeking out opportunities for growth, and building a supportive network of colleagues and mentors. By staying committed to their aspirations and leveraging their unique strengths, women can navigate gender-related challenges and make meaningful contributions to their field.

Over the years, there has been a gradual increase in gender diversity within engineering technology. While the field still has a long way to go, the presence of more women in technical roles is a positive trend. Meher observes that women tend to gravitate toward managerial roles, but there is a growing need for female representation in technical leadership positions, such as chief engineers.

Work–Life Balance

Maintaining a healthy work–life balance is crucial for long-term success and well-being, especially in demanding fields like engineering technology. For many professionals, including Meher, achieving this balance requires conscious effort, self-awareness, and the implementation of effective strategies.

In the early stages of her career, Meher found it challenging to maintain a healthy work–life balance. As a young professional eager to make a name for herself and keep up with her peers, she often prioritized work over personal time. Living alone without pets or significant others, Meher felt the pressure to dedicate long hours to her job, including working on weekends and designated off days. This intense focus on work, while initially seeming productive, eventually led to burnout and a realization that such a pace was unsustainable.

The turning point for Meher came when she began to experience the negative effects of her work-centric lifestyle. Burnout, characterized by physical and emotional exhaustion, decreased productivity, and a lack of motivation, made it clear that she needed to reassess her approach to work and personal life. Meher realized that maintaining a healthy work–life balance was not just about managing time but also about preserving her well-being and long-term career sustainability.

To address the imbalance, Meher began to implement several strategies aimed at creating a more sustainable and fulfilling work–life balance. She made a conscious decision to respect her designated off days, such as taking every other Friday off. Initially, she struggled with the temptation to work on these days, but she gradually learned to prioritize rest and relaxation. By taking these breaks seriously, she found that she returned to work more refreshed and productive.

Establishing clear boundaries between work and personal time was essential for Meher. She set specific work hours and made a commitment to stop working at a certain time each day. This helped her create a routine that allowed for personal activities and downtime, reducing the risk of burnout. Meher shifted her focus from the quantity of hours worked to the quality of work produced. She aimed to be more efficient and effective during her work hours, ensuring that she accomplished her tasks without the need for excessive overtime. This approach not only improved her productivity but also allowed her to enjoy her personal time without guilt.

To achieve a better work–life balance, Meher made time for activities that brought her joy and relaxation. Whether it was pursuing hobbies, spending time with friends and family, or simply unwinding with a good book, these activities helped her recharge and maintain a positive outlook. Meher sought advice from colleagues and mentors who had successfully navigated similar challenges. Their insights and experiences provided valuable guidance on how to manage work demands while maintaining a healthy personal life. This support network also offered encouragement and accountability as she worked to improve her work–life balance.

The benefits of Meher's efforts to achieve a healthier work–life balance were significant. By taking regular breaks and setting boundaries, she experienced increased productivity, improved mental and physical health, and greater overall job satisfaction. Her ability to manage stress and avoid burnout allowed her to bring her best self to both her professional and personal life. Moreover, Meher's commitment to work–life balance set a positive example for her colleagues and mentees. By demonstrating that it is possible to achieve professional success without sacrificing personal well-being, she encouraged others to prioritize their health and happiness. This, in turn, contributed to a more supportive and balanced work environment for everyone.

Achieving and maintaining a healthy work–life balance is an ongoing process that requires continuous effort and adjustment. Meher remains committed to evaluating her work habits and making necessary changes to ensure that she maintains a sustainable and fulfilling balance. She recognizes that life circumstances and work demands can change, and she is prepared to adapt her strategies as needed.

In conclusion, Meher's journey to achieving a healthy work–life balance highlights the importance of self-awareness, setting boundaries, and prioritizing well-being. By implementing effective strategies and seeking support, she has been able to create a more sustainable and fulfilling professional and personal life. Her experiences serve as a valuable guide for others in engineering technology who are striving to achieve a similar balance.

Advice for Aspiring Women Engineers

One of the first steps in choosing a career path is understanding your learning style and preferences. Engineering technology, with its emphasis on practical applications and hands-on learning, may be more suitable for those who prefer to see and do rather than solely engage with theoretical concepts. Meher, for instance, found that her hands-on, visual learning style was better suited to engineering technology than traditional engineering programs. Recognizing this early on can help you select a program that aligns with your strengths and interests.

Embracing Resilience and Persistence

The journey to becoming an engineer is often filled with challenges and setbacks. Meher's experiences highlight the importance of resilience and persistence. Whether it is facing rejection from a preferred program, struggling with coursework, or encountering gender-related biases, it is crucial to keep pushing forward. Embrace setbacks as learning opportunities and remain committed to your goals. Persistence and a positive mindset can help you navigate obstacles and achieve success.

Seeking Support and Mentorship

Building a strong support network is essential for personal and professional growth. Seek out mentors who can provide guidance, share their experiences, and offer valuable advice. Meher benefited greatly from the support of her professors and peers, who encouraged her and helped her navigate the complexities of her academic and professional journey. Do not hesitate to reach out to potential mentors, join professional organizations, and participate in networking events to build connections and gain insights from others in the field.

Developing Soft Skills

While technical expertise is crucial, soft skills such as communication, teamwork, and leadership are equally important. Meher emphasizes the value of effective communication and the ability to work well in teams. These skills are essential for collaborating with colleagues, presenting ideas, and navigating complex projects. Invest time in developing your soft skills through workshops, courses, and practical experiences. Strong interpersonal skills can set you apart and enhance your ability to contribute to your team and organization.

Balancing Assertiveness and Collaboration

In competitive environments, it is important to assert yourself and advocate for your ideas. However, this should be balanced with a collaborative approach that values the contributions of others. Meher's strategy involves speaking out confidently and respectfully, ensuring that her voice is heard while maintaining positive relationships with her colleagues. Practice assertive communication techniques and seek feedback to improve your ability to navigate team dynamics effectively.

Staying Updated with Trends and Technologies

Engineering technology is a rapidly evolving field, and staying updated with the latest trends and advancements is crucial. Meher leverages social media platforms like LinkedIn for professional updates and participates in internal technical career fairs and lunch-and-learn sessions. Engage with industry news, attend conferences, and participate in continuous learning opportunities to keep your skills and knowledge current. Staying informed about emerging technologies and best practices can enhance your expertise and career prospects.

Prioritizing Work–Life Balance

Maintaining a healthy work–life balance is essential for long-term success and well-being. Meher's journey highlights the importance of setting boundaries, taking breaks, and prioritizing personal activities. Recognize the signs of burnout and take proactive steps to manage stress and maintain your well-being. A balanced approach to work and personal life can improve your productivity, job satisfaction, and overall quality of life.

Advocating for Gender Diversity

As a woman in engineering technology, you have the opportunity to advocate for gender diversity and support other women in the field. Meher actively participates in initiatives aimed at empowering women engineers and shares her experiences to inspire the next generation. Get involved in professional organizations, mentorship programs, and advocacy efforts to contribute to a more inclusive and diverse engineering community. Your efforts can help create a supportive environment where all engineers can thrive.

Setting Clear Goals and Aspirations

Having clear career goals and aspirations can provide direction and motivation. Meher's journey was guided by her passion for engineering and her desire to make meaningful contributions to her field. Set specific, achievable goals and create a plan to reach them. Regularly assess your progress and adjust your goals as needed. Having a clear vision of what you want to achieve can help you stay focused and motivated throughout your career.

Embracing Lifelong Learning

Engineering technology is a field that requires continuous learning and adaptation. Embrace a mindset of lifelong learning and seek opportunities to expand your knowledge and skills. Whether it is pursuing advanced degrees, obtaining certifications, or participating in professional development programs, continuous learning can enhance your expertise and open new career opportunities. Stay curious, seek out new challenges, and remain committed to your growth and development.

In conclusion, aspiring women engineers can navigate the path to success by understanding their learning style, embracing resilience, seeking support, developing soft skills, balancing assertiveness with collaboration, staying updated with trends, prioritizing work–life balance, advocating for gender diversity, setting clear goals, and embracing lifelong learning. Meher's experiences and insights provide valuable guidance for young women considering a career in engineering technology, offering practical advice and inspiration for overcoming challenges and achieving their aspirations.

Staying Updated with Trends

In the rapidly evolving field of engineering technology, staying updated with the latest trends and advancements is crucial for maintaining relevance and enhancing career prospects. Continuous learning and adaptation are essential for engineers to keep pace with emerging technologies, industry best practices, and innovative solutions. Meher's approach to staying informed provides valuable insights for aspiring engineers.

Leveraging Social Media and Professional Networks

Social media platforms, particularly LinkedIn, play a significant role in keeping professionals updated with industry trends. LinkedIn offers a wealth of information through articles, posts, and discussions shared by industry experts, companies, and professional organizations. Meher actively engages with LinkedIn to follow thought leaders, join relevant groups, and participate in discussions. This allows her to stay informed about the latest developments, gain insights from experts, and network with professionals in her field.

Participating in Conferences and Workshops

Attending conferences, workshops, and seminars is another effective way to stay updated with trends. These events provide opportunities to learn about cutting-edge technologies, research findings, and industry innovations. Meher regularly attends engineering conferences and workshops, where she can interact with experts, attend presentations, and participate in hands-on sessions. These experiences not only enhance her knowledge but also provide valuable networking opportunities.

Engaging in Internal Technical Career Fairs

Many organizations, including Northrop Grumman, host internal technical career fairs and lunch-and-learn sessions. These events bring together experts from various fields to share their knowledge and experiences. Meher actively participates in these events, where she can learn about new technologies, best practices, and ongoing projects within her organization. Engaging with colleagues and experts in these settings helps her stay informed and connected with the latest advancements.

Subscribing to Industry Publications

Subscribing to industry publications, journals, and newsletters is a valuable way to receive regular updates on trends and developments. Meher subscribes to several engineering and technology journals, which provide in-depth articles, research papers, and case studies on emerging technologies and industry practices. These publications offer a comprehensive view of the latest advancements and help Meher stay informed about the broader trends shaping the field.

Online Courses and Certifications

Online courses and certifications offer flexible and accessible opportunities for continuous learning. Platforms like Coursera, edX, and Udacity provide courses on various engineering and technology topics, allowing professionals to expand their knowledge and skills. Meher takes advantage of these platforms to enroll in courses that align with her interests and career goals. Obtaining certifications in specialized areas can also enhance her expertise and credibility in the field.

Networking with Industry Experts

Building and maintaining a network of industry experts is crucial for staying updated with trends. Meher actively seeks out opportunities to connect with professionals who have expertise in her areas of interest. Networking can occur through professional organizations, conferences, social media, and mentorship programs.

Engaging with experts allows Meher to gain insights, ask questions, and stay informed about the latest developments in engineering technology.

Participating in Research and Development Projects

Involvement in research and development (R&D) projects provides hands-on experience with emerging technologies and innovative solutions. Meher's participation in R&D projects, such as her work on renewable energy systems at a wastewater plant, allows her to explore new concepts and contribute to advancements in the field. Engaging in R&D projects helps her stay at the forefront of technological innovation and apply her knowledge to real-world challenges.

Continuous Improvement and Adaptation

Staying updated with trends requires a commitment to continuous improvement and adaptation. Meher regularly evaluates her skills and knowledge, identifying areas for growth and seeking opportunities to expand her expertise. She remains open to new ideas and approaches, adapting her strategies to stay relevant in a rapidly changing field. This mindset of lifelong learning and continuous improvement is essential for success in engineering technology.

Engaging with Professional Organizations

Professional organizations, such as the Institute of Electrical and Electronics Engineers (IEEE) and the American Society of Mechanical Engineers (ASME), offer valuable resources for staying updated with trends. Meher is an active member of several professional organizations, where she can access publications, attend events, and participate in discussions. These organizations provide a platform for networking, learning, and professional development.

Utilizing Technology and Tools

Technology and tools, such as software applications and digital platforms, play a crucial role in staying updated with trends. Meher utilizes various tools to access information, collaborate with colleagues, and manage her learning. For example, she uses MBSE tools like Cameo from CATIA to stay informed about advancements in model-based systems engineering. Leveraging technology helps her stay connected and informed in a dynamic and fast-paced field.

In conclusion, staying updated with trends in engineering technology requires a proactive approach that includes leveraging social media, participating in conferences and workshops, engaging in internal technical career fairs, subscribing to industry publications, enrolling in online courses, networking with experts, participating in R&D projects, committing to continuous improvement, engaging with professional organizations, and utilizing technology and tools. Meher's strategies provide valuable guidance for aspiring engineers seeking to stay informed and enhance their career prospects in a rapidly evolving field.

Core Skills for Success

Achieving success in engineering technology requires a combination of technical expertise and soft skills. These core skills are essential for making meaningful contributions, advancing in one's career, and fostering a collaborative and innovative work environment. Meher's experiences highlight the importance of developing both technical and interpersonal skills to navigate the complexities of the field.

Technical Expertise

Technical expertise forms the foundation of a successful career in engineering technology. This includes a deep understanding of engineering principles, proficiency in relevant tools and technologies, and the ability to apply theoretical knowledge to practical problems. Meher's journey underscores the importance of continuously expanding one's technical skills through education, hands-on experience, and ongoing learning.

A strong grasp of engineering principles is crucial for solving complex problems and designing effective solutions. This includes knowledge of mechanics, materials science, electrical systems, and software engineering. Meher's education in mechanical engineering technology provided her with a solid foundation in these areas, enabling her to tackle diverse challenges in her role as a systems engineer.

Proficiency in using various tools and technologies is essential for engineers. This includes software applications for modeling, simulation, and analysis, as well as hardware tools for testing and prototyping. Meher's expertise in model-based systems engineering (MBSE) tools, such as Cameo from CATIA, allows her to create detailed digital models and improve project efficiency.

The ability to apply theoretical knowledge to real-world problems is a key skill for engineers. Meher's work on renewable energy systems at a wastewater plant demonstrates her ability to integrate technical expertise with practical applications.

This skill is essential for developing innovative solutions that address real-world challenges.

Problem-Solving and Critical Thinking

Problem-solving and critical thinking are core skills that enable engineers to analyze complex issues, identify root causes, and develop effective solutions. These skills involve a systematic approach to understanding problems, evaluating options, and implementing solutions.

Analytical skills are essential for breaking down complex problems into manageable components. Meher's ability to identify gaps and inconsistencies in system integration processes showcases her strong analytical skills. Engineers must be able to assess data, interpret results, and make informed decisions based on their analysis.

Creativity and innovation are crucial for developing novel solutions to engineering challenges. Meher's innovative approach to integrating renewable energy systems highlights her ability to think outside the box and explore new possibilities. Engineers must be open to new ideas and willing to experiment with different approaches to find the best solutions.

Effective decision-making involves evaluating options, considering potential risks, and choosing the best course of action. Meher's role as a systems engineer requires her to make critical decisions that impact project outcomes. Engineers must be able to weigh the pros and cons of different options and make decisions that align with project goals and customer requirements.

Effective Communication

Effective communication is essential for collaborating with colleagues, presenting ideas, and ensuring that project requirements are clearly understood. Meher's ability to articulate technical details in a way that is accessible to non-engineers is a key strength that enhances her effectiveness in her role.

Verbal communication skills are crucial for conveying ideas, providing updates, and facilitating discussions. Meher's ability to speak confidently and respectfully in meetings ensures that her contributions are recognized and valued. Engineers must be able to explain complex concepts clearly and engage in productive conversations with their teams.

Written communication skills are important for documenting processes, creating reports, and communicating with stakeholders. Meher's proficiency in writing detailed project reports and technical documentation helps ensure that information is accurately conveyed and easily understood. Engineers must be able to write clearly and concisely, providing essential information in a structured format.

Interpersonal skills involve building positive relationships, collaborating effectively, and navigating team dynamics. Meher's empathetic approach and ability to read the room contribute to a collaborative work environment. Engineers must be able to work well with others, listen actively, and foster a culture of teamwork and mutual respect.

Leadership and Teamwork

Leadership and teamwork are essential for guiding projects, motivating teams, and achieving collective goals. Meher's leadership style emphasizes inclusivity and collaboration, creating a positive and productive atmosphere.

Leadership skills involve inspiring and guiding teams, setting clear goals, and providing support and direction. Meher's role as a deputy systems engineering integration and test (SEIT) lead requires her to lead teams, ensure project alignment, and address challenges. Engineers must be able to lead by example, communicate vision, and empower their teams to succeed.

Teamwork involves working collaboratively with colleagues, sharing knowledge, and contributing to collective efforts. Meher's ability to engage with quieter team members and encourage their input fosters a cohesive and inclusive team environment. Engineers must be able to work effectively in teams, leveraging diverse perspectives and skills to achieve project goals.

Adaptability and Continuous Learning

Adaptability and continuous learning are crucial for staying relevant in a rapidly evolving field. Meher's commitment to lifelong learning and continuous improvement ensures that she remains at the forefront of technological advancements.

Adaptability involves being open to change, embracing new technologies, and adjusting strategies as needed. Meher's willingness to explore new approaches and adapt her methods to changing circumstances demonstrates her adaptability. Engineers must be able to navigate uncertainty, remain flexible, and respond effectively to new challenges.

Continuous learning involves seeking opportunities to expand knowledge, acquire new skills, and stay updated with industry trends. Meher engages in online courses, attends conferences, and participates in professional development programs to enhance her expertise. Engineers must be committed to lifelong learning, staying curious, and continuously improving their skills.

In conclusion, achieving success in engineering technology requires a combination of technical expertise, problem-solving and critical thinking, communication skills, leadership and teamwork, and adaptability and continuous learning. Meher's experiences highlight the importance of developing these core skills to navigate the complexities of the field, make meaningful contributions, and advance in one's career. Aspiring engineers can benefit from focusing on these skills to achieve their goals and thrive in the dynamic world of engineering technology.

Future Aspirations

Looking ahead, Meher envisions a career that continues to blend her technical expertise with leadership and mentorship roles. Her aspirations are driven by a desire to make meaningful contributions to the field of engineering technology, foster innovation, and support the next generation of engineers.

Technical Leadership

One of Meher's primary goals is to advance into technical leadership positions, such as becoming a chief engineer. In this role, she aims to guide and oversee complex engineering projects, ensuring that they meet technical standards and align with organizational goals. As a chief engineer, Meher would be responsible for setting the technical direction of projects, making critical decisions, and providing strategic guidance to engineering teams.

Her experience as a systems engineer has equipped her with the skills needed for technical leadership. She has developed a deep understanding of various engineering disciplines, honed her problem-solving abilities, and demonstrated her capacity to manage and integrate diverse technical components. Meher's ability to see the big picture while paying attention to detail makes her well-suited for a leadership role that requires both strategic vision and technical acumen.

Mentorship and Advocacy

Mentorship is another key aspect of Meher's future aspirations. She is passionate about supporting and empowering young engineers, particularly women and underrepresented groups in the field. Meher currently serves in formal mentorship roles, where she can provide guidance, share her experiences, and help mentees navigate the challenges of their careers.

Her commitment to mentorship extends beyond one-on-one relationships. Meher aims to advocate for initiatives and programs that promote diversity and inclusion in engineering. By participating in professional organizations, attending conferences, and engaging in public speaking, she hopes to raise awareness about the importance of gender diversity and create opportunities for aspiring engineers to thrive.

Continuous Learning and Innovation

Meher's dedication to continuous learning and innovation is central to her future aspirations. She recognizes that the field of engineering technology is constantly evolving, and staying at the forefront of advancements is essential for success. Meher plans to pursue further education and certifications to deepen her expertise in emerging technologies and best practices.

She is particularly interested in exploring areas such as renewable energy, sustainable engineering, and advanced manufacturing. By staying informed about the latest trends and developments, Meher aims to contribute to innovative solutions that address global challenges. Her commitment to lifelong learning ensures that she remains adaptable and ready to tackle new and complex problems.

Balancing Technical and Business Acumen

In addition to her technical aspirations, Meher is keen on developing her business acumen. She understands that successful engineering projects require not only technical excellence but also a strong understanding of business principles. During her graduate studies, Meher had taken courses in business management, finance, and project management to enhance her ability to lead projects that are both technically sound and financially viable. Additionally, Meher seeks to further her knowledge in the business and finance fields to pursue continuous learning of effective business and financial strategies.

Her interest in the intersection of engineering and business is reflected in her previous work on renewable energy systems, where she conducted both technical and financial analyses. By strengthening her business skills, Meher aims to take on roles that involve strategic planning, resource management, and stakeholder engagement.

Personal and Professional Growth

Meher's future aspirations also include a focus on personal and professional growth. She is committed to maintaining a healthy work–life balance, recognizing that well-being is essential for long-term success. Meher plans to continue setting boundaries, taking breaks, and engaging in activities that bring her joy and fulfillment.

Professionally, she aims to seek out new challenges and opportunities that push her out of her comfort zone. Whether it is taking on a new project, learning a new skill, or stepping into a leadership role, Meher is dedicated to continuous improvement and growth. Her proactive approach to career development ensures that she remains motivated and ready to seize new opportunities.

In conclusion, Meher's future aspirations are driven by a desire to advance into technical leadership, support and mentor young engineers, stay at the forefront of innovation, balance technical and business acumen, and focus on personal and professional growth. Her vision for the future is one of continuous learning, meaningful contributions, and a commitment to fostering a diverse and inclusive engineering community. Aspiring engineers can draw inspiration from Meher's goals and dedication as they navigate their own career paths in the dynamic field of engineering technology.

Conclusion

The journey of women in engineering technology is marked by resilience, innovation, and a commitment to excellence. Through the experiences and insights shared by Meher, we gain a deeper understanding of the unique challenges and contributions of women in this dynamic field. Her story serves as both an inspiration and a guide for aspiring engineers, highlighting the importance of perseverance, continuous learning, and the pursuit of meaningful goals.

Meher's educational journey illustrates the transformative power of hands-on learning and practical applications. Her early fascination with roller coasters and subsequent academic pursuits in mechanical engineering technology demonstrate the impact of nurturing curiosity and passion. Despite facing initial setbacks and academic challenges, Meher's resilience and adaptability enabled her to find success and thrive in her chosen field.

In her professional role as a systems engineer, Meher exemplifies the critical importance of integrating diverse engineering disciplines to achieve cohesive project outcomes. Her ability to identify gaps, facilitate communication, and implement innovative solutions underscores the value of systems engineering in advancing technology. Meher's contributions extend beyond technical expertise, as she actively mentors junior engineers and advocates for gender diversity in the workplace.

Women in engineering technology bring unique insights and perspectives that enhance team dynamics and project outcomes. Meher's empathetic approach, effective communication skills, and inclusive collaboration foster a positive and productive work environment. Her experiences highlight the importance of emotional intelligence, leadership, and teamwork in achieving success.

Despite the significant progress in increasing gender diversity, women in engineering technology continue to face gender-related challenges. Meher's journey reveals the need for assertiveness, resilience, and strategic approaches to navigate biases and systemic barriers. Her advocacy for diversity and inclusion, coupled with her mentorship efforts, contributes to creating a more equitable and supportive engineering community.

Maintaining a healthy work–life balance is essential for long-term success and well-being. Meher's experiences underscore the importance of setting boundaries, prioritizing personal activities, and seeking support to manage stress and avoid burnout. Her commitment to work–life balance serves as a valuable example for others in the field.

Aspiring women engineers can draw inspiration from Meher's advice, which emphasizes understanding one's learning style, seeking mentorship, developing soft skills, staying updated with trends, and advocating for gender diversity. Her proactive approach to career development and continuous learning ensures that she remains at the forefront of technological advancements.

Meher's future aspirations include advancing into technical leadership, supporting and mentoring young engineers, contributing to sustainable solutions, and balancing technical and business acumen. Her vision for the future is one of continuous learning, meaningful contributions, and a commitment to fostering a diverse and inclusive engineering community.

In conclusion, Meher's experiences and insights provide a comprehensive and inspiring view of women's journeys in engineering technology. Her story highlights the importance of resilience, innovation, and the pursuit of excellence. By embracing these values and focusing on core skills for success, aspiring engineers can navigate the complexities of the field, make meaningful contributions, and achieve their goals. The future of engineering technology is bright, and women like Meher are paving the way for a more diverse, inclusive, and innovative industry.

The Journey of an Aspiring Engineer

Olivia Engavall

The Early Spark

Olivia Engvall's journey into the world of engineering began in high school, where she joined a FIRST Robotics team during her freshman year. This experience was pivotal, as it provided her with a comprehensive introduction to the field of engineering. The collaborative environment of FIRST Robotics allowed Olivia to work closely with her peers and mentors, fostering a sense of teamwork and learning. She had the opportunity to explore programming, marketing, machine design, electrical, and mechanical fabrication. She was drawn to the mechanical and electrical systems and showed an aptitude for brainstorming designs and troubleshooting the robot.

Her hands-on involvement in these projects laid a strong foundation for her future endeavors. The skills she developed in high school—such as critical thinking, creativity, and technical skills—proved invaluable as she progressed in her educational journey. Olivia's background with FIRST Robotics not only solidified her passion for engineering but also guided her educational choices, leading her to pursue a degree in Mechanical Engineering Technology at Purdue University. This early exposure to the practical applications of engineering principles set the stage for her continued success and growth in the field.

O. Engavall (✉)
Purdue University, West Lafayette, IN, USA

A. Lucietto (ed.), *Innovating for Tomorrow: Engineering Technology*, Women in Engineering and Science, https://doi.org/10.1007/978-3-032-07645-8_19

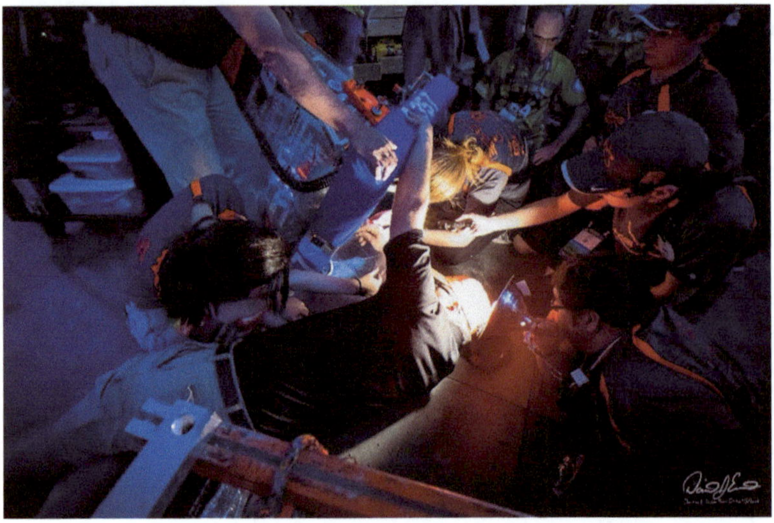

Olivia repairing her team's robot during a competition

Olivia using an angle grinder

The College Decision

When it came time to apply for college, Olivia felt mechanical engineering was a natural fit. Her passion for engineering, nurtured through her high school robotics experience, led her to Purdue University's Polytechnic Institute. Their School of Engineering Technology offered a unique blend of theoretical knowledge and practical application. The hands-on focus of the program was a significant factor in her decision, as it allowed her to continue the type of experiential learning she had enjoyed in high school. Unlike other engineering programs that start with a general first-year curriculum, Purdue's Polytechnic Institute starts students directly in a specialized degree. Olivia appreciated the energetic and practical approach of the Polytechnic Institute. The program's emphasis on real-world applications and problem-solving resonated with her and aligned with her learning style and interests.

A Year in Peru

While pursuing her degree in Mechanical Engineering Technology, Olivia decided to add a minor in Spanish. She had begun studying the language in high school and was interested in the opportunities to travel and improve her communication skills. Additionally, during an internship, the vice president of engineering had imparted upon all interns the value of intercultural education in an increasingly globalized world.

This decision opened up new opportunities for Olivia, including the chance to study abroad. In her final year of college, she chose to participate in a new Dual Degree Program between Purdue University and the Universidad de Ingeniería y Tecnología (UTEC) in Peru. Students from UTEC had completed the program at Purdue, but Olivia was the first representative from Purdue University to travel to Peru and complete the program.

Olivia was interested in the opportunity to improve her language skills and live abroad. This immersive experience allowed her to fully integrate into the academic and cultural environment of Peru, providing her with a unique perspective on engineering education and practice in a different country.

Olivia's time in Peru was marked by significant academic and personal growth. Immersed in a new cultural and educational environment, she embraced the opportunity to broaden her horizons. She took a variety of courses that complemented her studies at Purdue. She engaged in general education classes such as ethics and contemporary history of Peru, as well as mechanical engineering courses like machine design and finite element analysis.

Taking courses in another country and in another language added difficulty. Olivia was exposed to new areas of study such as mine engineering which is one of the top industries in Peru. Even courses that covered material she was familiar with included new challenges. For example, her course on machine design was impacted by the country's material availability. Although the textbook referenced many steel alloys, Peru only imports a few varieties which impacts the choices an engineer can make. Additionally, because the coursework was in Spanish, it forced her to study harder and become more organized.

One of the highlights of her education abroad was working on a thesis project with a partner. They focused on designing a soft robot gripper capable of harvesting blueberries, a project that combined the agricultural industry with the field of soft robotics. This project allowed Olivia to apply her engineering skills in a practical context, further growing her understanding of mechanical engineering principles.

The project involved extensive research on the agriculture in Peru, including the specific requirements for harvesting blueberries. Olivia and her partner had to consider various factors such as the delicacy of the fruit, the need for precision in the gripping mechanism, and the motion required for picking blueberries. They developed a CAD model and completed motion studies for the gripper. This included detailed design specifications and demonstrated its effectiveness. The project not only enhanced Olivia's technical skills but also taught her important lessons in research, teamwork, and problem-solving.

Living in Peru also presented its own set of challenges and adventures. Olivia had to adapt to a new language, which required her to practice and improve her Spanish skills continuously. This linguistic immersion was both challenging and rewarding, as it enhanced her ability to communicate effectively and deepened her understanding of the local culture. Olivia visited the Amazon Rainforest and villages in the Andes, and between her two semesters, she took a month-long journey though the southern half of the country to see Lake Titicaca and hike to Machu Pichu.

Overall, Olivia's decision to add a minor in Spanish and study abroad in Peru was transformative. It provided her with a deeper understanding of mechanical engineering, exposed her to new cultural perspectives, and helped her develop both personally and professionally. The challenges she faced and the adventures she embarked on helped shape her into a more independent, culturally aware, and resourceful individual.

Olivia at Machu Pichu

Olivia with her cohort of international students at UTEC

Overcoming Challenges

Throughout her educational journey, Olivia faced several challenges that tested her resilience and determination. In her first year of study, she experienced some anxieties about whether she had chosen the right degree field, as some industries view engineering technology as more of a high-level maintenance role rather than a full-fledged engineering discipline. Olivia worried about potential limitations in her career prospects and whether her degree would be respected in the professional world.

However, with the support of her advisors and professors, Olivia gradually gained confidence in her choice. Her mentors offered insights into the strengths of the engineering technology program, emphasizing its practical, hands-on approach. They reassured her that the skills she was learning were highly relevant and sought after in many industries. Olivia realized that the hands-on practice provided by her program was invaluable and well-suited to her interests. This practical focus allowed her to develop real-world skills that would be directly applicable in her future career, ultimately alleviating her concerns.

Another challenge Olivia faced was being out of state and not knowing anyone at Purdue. This situation initially felt daunting, as she had to navigate a new environment without the comfort of familiar faces. However, this challenge turned into a strength, as it forced her to make independent decisions about her classes and extracurricular activities. Olivia had the freedom to explore her interests and forge her own path independent of peer influence from her high school friends. This helped her grow both personally and academically, as she learned to rely on her own judgment and take initiative in her educational journey.

The COVID-19 pandemic added another layer of difficulty to Olivia's educational journey. The sudden shift to remote learning and the uncertainty surrounding the pandemic added stress. However, Olivia's extensive internship experience provided her with a real-world perspective that kept her motivated. These internships allowed her to see the direct impact of her work and keep a sense of purpose despite the challenges posed by the pandemic.

Studying abroad in Peru presented unique challenges as well. Olivia had to adapt to learning in a second language, which required her to spend more time on reading, writing, and double-checking her work. It was demanding, but language immersion pushed her to develop her language skills to a higher level. The experience honed her focus and study skills, pushing her to new levels of academic discipline. Olivia's perseverance in overcoming these language barriers not only enhanced her communication abilities but also deepened her cultural understanding and appreciation.

Professional Growth

One of the pivotal moments during Olivia's education was securing her first internship. The Vice President of Engineering at Amway visited her high school robotics team and was impressed by the students' level of skill and dedication. This visit led to the decision to offer a part-time internship, initially termed a co-op, to Olivia and another student. This opportunity allowed them to work on smaller projects and research-based tasks, providing them with valuable real-world practice.

During this co-op, Olivia and her fellow intern were given a variety of assignments that ranged from conducting research to assisting with minor engineering projects. Although the program was experimental and did not continue beyond their involvement, their performance was noteworthy, leading to further invitations for internships at Amway.

Olivia completed three summers of internships with Amway, each building on the knowledge and skills she had gained. These internships were crucial in her professional development, offering her insights into the practical applications of engineering principles in a corporate setting. The internships included a mix of hands-on projects and virtual work, specifically during the COVID-19 pandemic. The virtual internship was unique and taught Olivia to use digital communication tools, skills she now uses today as it becomes more common to collaborate remotely.

After her internships with Amway, Olivia sought to explore other opportunities. She secured a summer internship with Caterpillar, where she gained exposure to manufacturing strategy and Industry 4.0. This internship provided her with a fresh perspective and additional skills that complemented her previous experiences. Following graduation, Olivia took an Operations Engineer position at MillerKnoll (formerly Herman Miller), where she supported manufacturing and continued to gain valuable industry exposure.

Ultimately, Olivia returned to Amway, drawn by the strong connections she had built during her internships and the company's supportive environment. Networking played a crucial role in her career development, as the relationships she had cultivated helped her secure positions at both MillerKnoll and Amway. The importance of networking was evident in her ability to leverage these connections to find opportunities and advance her career.

Olivia's journey with Amway, from her first internship to her current role, highlights the significance of internships and networking in shaping a successful engineering career. Her experiences underscore the value of building strong professional relationships and continuously seeking opportunities for growth and development.

Olivia and her cohort of engineering interns at Amway

Mentorship and Role Models

Throughout her journey, Olivia benefited immensely from the guidance of mentors and role models who played a crucial role in her development as an engineer. In high school, her robotics team mentors, who were professional engineers, provided invaluable support and encouragement. These mentors not only shared their technical expertise but also inspired Olivia with their passion for their carriers. They guided her through complex projects, taught her problem-solving techniques, and helped her develop a solid foundation in mechanical engineering. Their mentorship was instrumental in shaping Olivia's interest and aptitude for the field.

At Purdue University, Olivia continued to receive support from several professors who were dedicated to her success. Faculty members provided guidance on coursework, research projects, and career planning. Their mentorship helped Olivia navigate the challenges of her degree program and gain confidence in her abilities. The professors' commitment to fostering a supportive learning environment enabled Olivia to thrive academically and personally.

Olivia also remained involved with FIRST Robotics as a mentor, giving back to the community that had been so influential in her own development. As a college mentor, she helped high school students by sharing knowledge and advice, providing guidance on robotics projects, and encouraging them to pursue their interests in engineering. This role allowed Olivia to continue learning from older mentors while simultaneously becoming a mentor herself. The reciprocal nature of this mentorship improved her communication and project management skills while reinforcing the importance of community and collaboration.

In her current role at Amway, she was initially assigned a "buddy" to help her integrate into the corporate environment. This relationship evolved into an official mentorship, with regular bi-weekly meetings to review company systems, processes, and problem-solving strategies. These sessions provided Olivia with a comprehensive understanding of the company's operations and equipped her with the tools needed to succeed in her role.

Additionally, Olivia was partnered with an experienced engineer who had previously held her position. This mentor guided her through the intricacies of her job, offering shadowing opportunities and insights into project management. The hands-on guidance and practical advice from this mentor were invaluable, helping Olivia to quickly adapt to her responsibilities and excel in her position.

The structured mentorship at Amway has been a cornerstone of Olivia's professional growth. The support and knowledge she received from her mentors have enabled her to tackle complex projects, develop innovative solutions, and continuously improve her skills. Olivia's journey underscores the profound impact that mentorship and role models can have on an individual's career, highlighting the importance of seeking guidance and building strong professional relationships.

Current Role and Responsibilities

Olivia was hired as a project engineer at Amway's Nutritional Products Plant, a facility dedicated to producing protein powders and drink mixes. Her role as a project engineer involves overseeing large-scale capital projects that are critical to the plant's operations and efficiency. These projects include purchasing new equipment, upgrading existing machinery, and implementing changes to enhance production capabilities. Olivia's responsibilities require her to manage the entire lifecycle of these projects, from planning to installation. Olivia's technical ability and project management skills are essential in successfully navigating these complex tasks.

Recently, Olivia has taken on added responsibilities as a hybrid project and process engineer. This expanded role allows her to support day-to-day operations while continuing to manage large-scale projects. As a process engineer, Olivia runs plant trials involved with product launches and works with the plant support team to ensure that production runs smoothly and efficiently. She works closely with the plant's technicians to find areas for improvement and implement solutions that enhance productivity and reduce downtime.

Olivia's ability to balance her responsibilities as both a project and plant engineer shows her versatility and commitment to continuous improvement. Her work on large-scale capital projects and day-to-day operations ensures that the plant is competitive and capable of meeting the evolving needs of the market. Through her efforts, Olivia contributes to the plant's success and helps drive innovation in the production of nutritional products.

Rewarding Projects

Olivia finds her work in manufacturing rewarding, primarily because of the direct impact it has on plant efficiency and the quality of work for plant technicians. Her role allows her to make tangible improvements that enhance work comfort and safety, as well as the overall productivity of the plant. Olivia takes pride in knowing that her efforts contribute to the smooth functioning of the production processes and the creation of high-quality nutritional products.

Alongside improving the quality of work, Olivia values that she can incorporate the insights and opinions of plant technicians into her projects. She recognizes that these employees, with their hands on experience of the plant's operations, offer invaluable perspectives. By actively seeking their input and collaborating with them, Olivia ensures that her projects are not only technically sound but also practical. This collaborative approach fosters a sense of ownership and engagement among the plant technicians, as they see their ideas and feedback being implemented to improve their work environment.

Olivia's ability to make these smaller improvements is a testament to her problem-solving skills and attention to detail. She approaches each task with a focus on practical solutions that can be quickly and effectively implemented. This mindset allows her to address immediate needs and make continuous progress, even as larger projects are still in development.

The satisfaction Olivia derives from her work is also tied to the positive relationships she has built with her colleagues. By valuing their expertise and fostering a collaborative atmosphere, she has created a supportive and dynamic work environment. This sense of community and shared purpose enhances the overall morale of the team and drives collective success.

Ongoing Challenges

Being early-in-career, the most common challenge Olivia faces is being unsure of what to do next. In a project or during a meeting, she is often aware of her lack of knowledge and how it affects her ability to make decisions quickly. These situations, while becoming less frequent over time, continue to push her to develop her problem-solving skills and rely on her mentors for guidance.

When faced with uncertainty, Olivia has learned to take a methodical approach to problem-solving. She begins by breaking down the issue into smaller, more manageable components. This allows her to find the next small task that needs to be completed. By focusing on one aspect at a time, Olivia can systematically work through the challenge without becoming overwhelmed.

This often involves gathering information and collaborating with her team. Olivia finds that being unsure is an indicator that she needs to reach out for guidance. She recognizes that effective problem-solving is rarely a solitary endeavor and that leveraging the collective knowledge and expertise of her colleagues can lead to more innovative and effective solutions. By involving her colleagues in the problem-solving process, Olivia not only finds effective solutions but also strengthens the bonds within her team.

In these moments of uncertainty, Olivia also relies heavily on her mentors. Their experience and insights provide her with valuable perspectives that she might not have considered on her own. Regular meetings with her mentors allow Olivia to discuss the challenges she is facing, receive constructive feedback, and explore different approaches to problem-solving. This mentorship not only helps her navigate immediate issues but also contributes to her long-term professional development.

Over time, Olivia has become more confident in her ability to handle complex challenges. While she still encounters situations where she feels unsure, her growing experience and the support of her mentors and team have equipped her with the tools to navigate these difficulties successfully. Each challenge she overcomes adds to her skill set and prepares her for future obstacles, reinforcing her resilience and contributing to her growth as a capable and confident engineer.

Perspective as a Woman in Engineering

Olivia has been fortunate to avoid the negative interactions so often shared by women in male-dominated fields. However, whether stemming from being a woman or being early-in-career, she has occasionally noticed a lack of respect. This often appears through a colleague second-guessing her or not taking her advice at face value.

To navigate these challenges, Olivia consciously works on maintaining her confidence. She often reminds herself to approach situations with self-assurance and assertiveness. This mindset helps her to stand her ground and assert her opinions, ensuring that her contributions are recognized and valued. Olivia emphasizes the importance of not deferring unnecessarily to others, especially in situations where she has the knowledge and skills to make informed decisions. By recognizing her own expertise and trusting her abilities, she can effectively advocate for herself and her ideas.

Respect and collaboration are central to Olivia's role. She strives to respect the knowledge of manufacturing workers, who may not have higher education but have valuable practical insights gained from years of hands-on experience. Olivia understands that these workers have a deep understanding of the machinery and processes they work with daily, and their input is crucial for successful project outcomes. She actively seeks out the opinions and suggestions of manufacturing workers, integrating their practical knowledge into her engineering solutions. By valuing their contributions, Olivia fosters a collaborative environment where everyone's contributions are appreciated. This approach not only leads to better outcomes but also strengthens the relationships within the team, creating a more inclusive and supportive workplace.

In summary, Olivia's experience as a woman in engineering highlights the importance of having confidence, recognizing one's own expertise, and respecting the practical knowledge of colleagues. Her collaborative approach to problem-solving ensures that diverse perspectives are considered, leading to better outcomes and a more inclusive work environment.

Advice for Young Women

Olivia is a passionate advocate for encouraging young women to pursue careers in engineering, emphasizing that many of the fears about the field are unfounded. She believes that creativity and a willingness to learn are the most critical attributes for success in engineering. Being an engineer does not always require one to be a technical expert in every area or to have an exceptionally strong foundation in math. While technical skills are important, she argues that problem-solving, creativity, and the ability to learn are even more crucial.

Olivia entered the powder manufacturing industry without prior specialized knowledge. She had to learn niche information through various resources, including articles, webinars, courses, and conferences. This process was challenging but also incredibly rewarding, as it allowed her to develop a deep understanding of a complex and specialized field. Olivia's journey shows that with determination and a proactive approach to learning, anyone has what it takes to succeed in engineering.

Olivia also highlights the importance of hands-on opportunities. She appreciates the practical focus of her education at Purdue, which is particularly beneficial in manufacturing. This practical approach prepares graduates to understand and work with equipment effectively, bridging the gap between theoretical knowledge and real-world application. Olivia believes that hands-on experience is crucial for developing the skills and confidence needed to excel in engineering roles. She believes this approach better prepares students for the demands of the engineering profession.

Olivia advises young women that if they are creative and enjoy solving problems, they have what it takes to thrive in engineering. She emphasizes that the field is about innovation and finding practical solutions to real-world challenges, from designing new products to improving manufacturing processes. Furthermore, Olivia stresses the importance of resilience and adaptability. Engineering projects often come with unexpected challenges, and the ability to navigate these obstacles with a positive attitude and a willingness to learn is essential.

Olivia also encourages young women to seek out mentorship and build strong networks. She believes that having supportive mentors and a community of peers can significantly enhance one's confidence and offer valuable guidance throughout their career. Networking allows individuals to learn from each other, gain insights into different fields, and find opportunities for growth and development.

Staying Current

To stay current on the latest developments in her field, Olivia takes a proactive approach to continuous learning. She subscribes to newsletters related to powder and bulk solids manufacturing, which provide her with regular updates on industry trends, technological advancements, and best practices. These newsletters are a valuable resource, offering insights from experts and keeping her informed about the latest research and innovations.

In addition to reading newsletters, Olivia attends webinars that cover a wide range of topics relevant to her work. These webinars are an excellent way for her to gain in-depth knowledge on specific subjects, learn from industry leaders, and stay updated on new methodologies and technologies. Olivia also plans to attend industry conferences and expos. These events offer a unique opportunity to network with other professionals, attend presentations, and explore the latest technologies in the field. By attending these events, Olivia can gain firsthand experience with new tools and techniques, as well as build connections with other experts and practitioners.

Within her company, Olivia has a network of developing powder experts. This group regularly shares articles, webinars, and learning opportunities, fostering a culture of continuous improvement and knowledge sharing. By staying connected with her peers, Olivia can discuss challenges, brainstorm solutions, and stay informed about the latest developments in powder manufacturing. This network is particularly valuable for addressing complex issues, as it allows Olivia to draw on the collective expertise of her colleagues.

By leveraging a variety of resources and engaging in continuous learning, Olivia ensures that she remains at the forefront of her field. This proactive approach not only enhances her own skills and knowledge but also contributes to the overall success and innovation of her team and company.

Olivia's role as a mentor allowed her to share her knowledge and background, inspiring the next generation of engineers. Currently as a judge for the competitions, Olivia finds great satisfaction in seeing the students' growth and enthusiasm, knowing that she is contributing to the program and encouraging students to pursue careers in STEM fields.

In addition to her work with FIRST Robotics, Olivia is involved in other initiatives that promote professional development and networking among women in engineering. She participates in Inform, a professional development group, which focuses on education and career advancement, specifically for women. This group provides a platform for young professionals to connect, share experiences, and support each other's career growth. Olivia values the opportunity to engage with other women in manufacturing and engineering, both inside and outside her company, and believes that these connections are crucial for fostering a supportive and inclusive professional community.

Overall, Olivia's passion for these initiatives and programs reflects her commitment to continuous learning, mentorship, and community engagement. She understands the value of giving back and is dedicated to supporting the growth and development of future engineers. Through her volunteer work and advocacy, Olivia contributes to a more dynamic and inclusive engineering community, ensuring that others have the opportunities and support they need to succeed.

Redefining the Blueprint—Mica Caddick's Path into Engineering Technology and Beyond

Mica Caddick

Mica Caddick's journey into the world of STEM is anything but conventional. It begins not in a lab or workshop but in a high school English classroom, where she once imagined herself shaping young minds through literature. Her early aspirations were rooted in storytelling, language, and the humanities—worlds far removed from torque calculations, CAD software, or manufacturing systems. Yet precisely this unexpected shift—from aspiring English teacher to accomplished engineering technology professional and K-6 educator—would define her story.

M. Caddick (✉)
Purdue University, West Lafayette, IN, USA
e-mail: mica.caddick@nwsc.k12.in.us

A. Lucietto (ed.), *Innovating for Tomorrow: Engineering Technology*, Women in Engineering and Science, https://doi.org/10.1007/978-3-032-07645-8_20

This chapter explores the pivotal transitions that shaped Mica's career and identity. Her path was not a straight line but a series of discoveries, challenges, and reinventions. A chance exposure to hands-on technical work—perhaps through a robotics club, a vocational program, or a mentor who saw potential beyond the page—sparked a curiosity that grew into a passion for applied problem-solving. Unlike traditional engineering, which often emphasizes theoretical design, engineering technology offered Mica a more tactile, immediate connection to the real world. It was a field where she could see the results of her work take shape in physical systems, processes, and solutions.

But the transition was not easy. Mica had to overcome the steep learning curve of technical coursework and the internal doubts that come with stepping into unfamiliar territory. As a woman entering a male-dominated field, she often navigated spaces where she was underestimated or overlooked. Yet, these moments became catalysts for growth. With the support of mentors, peers, and educators who believed in her potential, Mica began to thrive—finding confidence in her skills and clarity in her purpose.

Her experiences in the field of engineering technology were as diverse as they were formative. From troubleshooting mechanical systems to optimizing production workflows, Mica developed a deep appreciation for the practical, problem-solving nature of her work. These real-world applications not only honed her technical abilities but also revealed a new calling: education. She realized that her journey—marked by detours, resilience, and discovery—could serve as a powerful example for others.

Now a STEM educator in Indiana, Mica brings her full self into the classroom: the former English enthusiast, the hands-on technologist, the mentor, and the advocate. She teaches not just equations and systems but the value of persistence, adaptability, and believing in one's ability to grow. Her story is a testament to the idea that STEM is not a fixed path but a dynamic landscape—one where curiosity can lead to transformation and where every background has a place.

A Shift in Trajectory: From Libraries to Labs

Mica Caddick's initial career aspirations were rooted in the quiet, contemplative world of books and learning. She envisioned herself immersed in literature, guiding students through the intricacies of language, and eventually pursuing a master's degree in library science. Her plan was clear, structured, and deeply aligned with her early academic strengths. Engineering, at that point, was not even a distant consideration.

That all changed during her senior year of high school.

What began as a casual involvement with her school's FIRST Robotics team quickly evolved into a transformative experience. Initially drawn in by the team's collaborative spirit and creative energy, Mica found herself captivated by the process of building, testing, and refining mechanical systems. The hands-on nature of the work—designing components, troubleshooting problems, and watching ideas come to life—sparked a curiosity she had not felt before. It was a different kind of learning: tactile, immediate, and deeply satisfying.

Her team's lead mentor, a local engineer and passionate advocate for STEM education, recognized her potential early on. He encouraged her to consider a future in engineering technology, emphasizing that her problem-solving mindset and attention to detail were valuable assets in the field. His belief in her abilities planted a seed of possibility that began to grow.

Encouraged but uncertain, Mica made a bold decision: she pivoted from her original plan and enrolled at Purdue University. The transition was not without its doubts. The coursework was rigorous, the environment unfamiliar, and the pressure to succeed was intense. At one point, overwhelmed and questioning her choice, she seriously considered leaving the program to attend trade school instead—a path she believed might offer more immediate, hands-on experience and a clearer sense of direction.

But two voices kept her grounded: her academic advisor, who reminded her of the long-term value of her education, and her father, who offered unwavering support and encouragement. They urged her to stay the course, to give herself time to adapt and grow. Their guidance proved pivotal.

Her eventual transfer into the MET program marked a turning point—not just academically but also personally. It was here that Mica began to truly fall in love with engineering technology. The program's emphasis on applied learning, real-world problem-solving, and industry-relevant skills resonated deeply with her. She thrived in labs and group projects, where she could see the tangible results of her work and collaborate with peers who shared her passion for innovation.

This period of transition—from aspiring librarian to emerging engineer—was more than a change in major. It was a redefinition of identity, a leap into the unknown, and the beginning of a journey that would shape her future in ways she never imagined.

Engineering Adventures—Building a Career from the Ground Up

Mica Caddick's time at Purdue University was not just about lectures and labs—it was a launchpad into the real world of engineering technology. As she progressed through the Mechanical Engineering Technology (MET) program, she began to see how the principles she was learning in the classroom translated directly into industry. Her education was grounded in application: troubleshooting systems, optimizing processes, and working with the kinds of tools and technologies used on the factory floor. It was this practical, career-focused approach that prepared her for the next phase of her journey.

One of the most formative experiences of her undergraduate years came through her internships with Chrysler, where she stepped into the fast-paced world of automotive manufacturing. These internships were more than résumé builders—they were immersive, hands-on experiences that tested her adaptability and deepened her technical skills. Mica quickly proved herself capable of handling complex challenges, earning the respect of her colleagues and mentors alike.

Upon graduation, Chrysler offered her a full-time position—a testament to her performance and potential. Over the next five years, Mica took on many roles within the company. She began as a quality manager, where she was responsible for ensuring that manufacturing processes met rigorous standards. Her role required a sharp eye for detail, a strong understanding of systems, and the ability to communicate effectively across departments.

But Mica did not stop there. She collaborated closely with the quality engineering team, diving into root cause analysis, process improvement, and data-driven decision-making. She also supported the resident engineering team, bridging the gap between design intent and production realities. She eventually found her way into the engineering department, and by the time she left she had been promoted to Resident Engineer. Her work spanned everything from supplier coordination to on-the-ground problem-solving—showcasing the versatility that engineering technologists are known for.

These years were intense, rewarding, and formative. Mica thrived in a dynamic environment, where no 2 days were the same, and every challenge was an opportunity to learn. Her ability to move fluidly between roles and responsibilities highlighted her technical competence and leadership potential.

That potential was recognized when she accepted a director of engineering position at an educational robotics company—a role that combined her technical expertise with her growing interest in mentorship and outreach. Here, Mica led teams, managed product development cycles, and contributed to the design of educational tools that would inspire the next generation of engineers. It was a dream role that aligned perfectly with her values and vision.

But in 2023, an unexpected job loss brought everything to a halt.

The sudden shift forced Mica into a period of deep reflection. She was not solving engineering problems or leading technical teams for the first time in years. Instead, she found herself substitute teaching at a local elementary school—an opportunity that initially felt like a temporary detour. She was hesitant, especially about working with young children, but necessity and curiosity led her to try it.

What she discovered surprised her.

In the classroom, Mica found a new kind of fulfillment. She saw the same spark of curiosity in her students that had once driven her to join a robotics team. She realized that her ability to explain complex ideas in simple, engaging ways was a gift—and that her journey through engineering technology had uniquely prepared her to teach. She accepted without hesitation when the school offered her a full-time position as their STEM and library teacher.

In this new role, Mica brought her story full circle. She merged her early love for literature and learning with her technical background, creating a classroom environment where students could explore both books and STEM topics. She designed hands-on STEM activities, introduced coding and robotics, and fostered a culture of curiosity and creativity. Her classroom became a space where students could see themselves as future engineers, scientists, and innovators—regardless of their background or starting point.

Mica's engineering adventures had taken her from the factory floor to the front of the classroom. And in doing so, she discovered that the most powerful thing an engineer can build is not just a product but a path for others to follow.

Building a STEM Program from the Ground Up

When Mica Caddick stepped into her new role as the first STEM teacher at her elementary school, she was not handed a curriculum, a guidebook, or a clear roadmap. Instead, she received a blank slate and the opportunity to design something transformative. Drawing on her background in engineering technology, where problem-solving, systems thinking, and hands-on application are core principles, Mica approached the challenge like an engineer: assessing the needs, designing a solution, testing, iterating, and improving.

Her responsibilities were vast and varied. She taught kindergarten through sixth-grade students, tailoring lessons to various developmental levels and learning styles. From introducing kindergartners to basic coding concepts using visual programming tools to guiding sixth graders through engineering design challenges, Mica ensured that every student had access to meaningful, age-appropriate STEM experiences.

But her role extended far beyond the classroom. She organized the school's science fair, transforming it from a modest event into a vibrant celebration of inquiry and innovation. She also revived the school's robotics team, allowing students to engage in competitive, collaborative engineering projects—an echo of her formative experience with FIRST Robotics. These initiatives were not just extracurriculars but extensions of her belief that STEM education should be immersive, inclusive, and inspiring.

Mica's commitment to curriculum development was relentless. She researched best practices, adapted engineering concepts for young learners, and aligned her lessons with state standards. Her engineering technology training gave her a unique advantage: She knew how to break down complex systems into understandable parts, and she applied that same logic to her teaching. Whether designing a unit on renewable energy, getting kindergarten students coding independently, or building a lesson around structural integrity using spaghetti and marshmallows, Mica made STEM tangible and exciting.

Her impact was not limited to STEM. As the school's library teacher, she also took on the challenge of revitalizing a space many students had seen as outdated or uninviting. She reorganized the collection, introduced interactive reading programs, and created cozy, engaging reading nooks. Her efforts paid off—students who once avoided the library began to visit eagerly, discovering genres and authors they loved. In many ways, Mica was engineering a curiosity culture that bridged the gap between books and bots.

Yet, the transition into teaching was not without its hurdles. Mica is simultaneously working toward becoming a fully licensed teacher, balancing her classroom responsibilities with coursework, certification requirements, and professional development. It was a demanding period, but she navigated with the same resilience and adaptability that had carried her through her engineering career.

Her master's degree in leadership and management, which she initially pursued to advance her career in industry, proved unexpectedly valuable in the classroom. It helped her manage time, lead projects, and build systems for student engagement and assessment. She applied organizational strategies to curriculum planning and used data-driven approaches to refine her teaching methods—just as she had done with engineering processes.

Mica's commitment to lifelong learning remained constant. She pursued additional certifications in STEM education and instructional design. She began exploring the possibility of a PhD in organizational psychology, a field that would allow her to study how people learn, lead, and grow within systems. Her curiosity had never waned; it had simply evolved.

In building a STEM program from scratch, Mica was not just teaching science and technology—she was modeling what it means to be an engineer in the truest sense: someone who sees a need, imagines a better way, and builds it. Her classroom became a living lab where students learned how to code or construct and think critically, collaborate, and believe in their ability to create change.

Mentoring: The Power of Being Seen and Supported

For Mica Caddick, mentoring has been both a lifeline and a legacy.

Her journey into engineering technology was sparked by a mentor—her FIRST Robotics coach—who saw potential in her before she saw it in herself. At a time when she was planning a future in English education, his encouragement nudged her toward a new path. He did not just teach her how to build a robot but he helped her build confidence, curiosity, and a sense of belonging in a field where she was often one of the few women in the room.

That early experience shaped Mica's understanding of what mentoring can do. It is not just about guidance—it is about recognition. It is about someone saying, *"You can do this. You belong here."*

As she advanced in her career—from launching new vehicle powertrains at Chrysler to leading engineering teams—Mica carried that lesson with her. She sought out mentors who challenged and supported her, and she became a mentor herself, especially to young women navigating the same uncertainties she once faced.

Her most fulfilling mentoring moments have come through FIRST Robotics, where she coached high school students to build championship-level robots. But more than the trophies, the transformation she witnessed in her students mattered most: the quiet student who found her voice as a team leader, the hesitant builder who became a confident designer. Mica did not just teach them engineering—she helped them see themselves as engineers.

Now, as a STEM educator, mentoring is woven into everything Mica does. She creates a classroom culture where questions are celebrated, mistakes are learning tools, and every student is encouraged to lead. She knows that for many of her students—especially girls and underrepresented minorities—she may be the first STEM role model they have ever had.

"Mentorship is about more than giving advice," *Mica says.* ***"It's about showing someone what's possible—and walking beside them until they believe it too."***

Her story reminds us that mentoring does not require perfection or a title. It requires presence, empathy, and the willingness to invest in someone else's potential. In doing so, mentors like Mica do not just shape careers—they change lives.

Reflections in the Rearview—Purpose, Perseverance, and the Power of Possibility

As Mica Caddick reflects on her multifaceted career, she sees not a single path but a series of interconnected journeys—each one shaped by curiosity, courage, and a commitment to growth. From launching vehicle platforms at Chrysler to launching young minds into the world of STEM, her story is one of transformation through action.

Among the many projects that stand out, a few remain especially close to her heart. Her work on vehicle platform launches at Chrysler was a defining experience—one that tested her technical skills, leadership, and ability to collaborate across disciplines. These were high-stakes, high-impact projects where precision mattered and timelines were tight. For Mica, seeing a powertrain roll off the production line and seeing the vehicle on the road, knowing she had contributed to its quality and performance, was a moment of profound pride. It was the kind of tangible, real-world outcome that engineering technologists are trained to deliver.

Equally meaningful was her initiative to work with a team to start a STEM camp for girls, a project born from her desire to create the kind of opportunity she wished she had as a young student. Together with other women from Chrysler, the camp provided a safe, supportive space for girls to explore robotics, coding, and engineering challenges—often for the first time. Watching their confidence grow as they built, tested, and presented their projects reminded Mica why representation and early exposure matter so deeply in STEM.

Her role as a FIRST Robotics coach brought her full circle. Guiding high school students through the process of designing and building championship-level robots was not just a technical endeavor—it was a mentorship mission. She saw herself in many of her students: unsure at first, but full of potential. The joy of watching them succeed, not just in competition but in discovering their own capabilities, was one of the most fulfilling chapters of her career.

Yet, Mica's journey has not been without adversity. As a woman in engineering, she often found herself navigating environments where she had to prove her competence repeatedly, sometimes in the face of skepticism or outright hostility. She encountered toxic workplace dynamics that tested her resilience and forced her to advocate for herself in ways that were both exhausting and empowering. These experiences, while painful, shaped her into a more empathetic leader and educator. They gave her a deeper understanding of the barriers underrepresented groups face in STEM and a fierce determination to help dismantle them.

In the classroom, Mica channels these lessons into action. Her teaching is infused with empathy, equity, and encouragement. She integrates emerging technologies, adapts her curriculum to reflect real-world applications, and creates a learning environment where every student feels seen and capable. Her students do not just learn about engineering—they learn to think like engineers: to ask questions, test ideas, and persist through failure.

Looking ahead, Mica envisions a future where she continues to evolve as both an educator and a learner. She is considering teaching part-time at the college level, where she can mentor aspiring engineers and technologists while continuing to shape K–12 education. Her passion for FIRST Robotics remains strong, and she hopes to expand her outreach to support more young women entering technical fields.

She is also contemplating further academic pursuits, including a Master's in Library Science and a potential PhD in organizational psychology. This field would allow her to explore how people learn, lead, and thrive within complex systems, a natural extension of her interests in leadership, education, and human development.

Mica Caddick's journey is a testament to the power of engineering technology not just to build machines but to build lives. Her story illustrates how applied knowledge can lead to profound personal and societal impact when paired with purpose and perseverance. She is an engineer, an educator, a mentor, and a changemaker, and her work continues to inspire those who dare to take the road less traveled.

Engineering Possibilities: Advice from Mica Caddick to Future Innovators

Mica Caddick did not always plan to become an engineer. Her journey began with a love for literature and a dream of becoming a librarian. But a spark of curiosity, ignited by a high school robotics team, led her down a path she never expected—into engineering technology, where she would build cars, lead teams, and eventually inspire the next generation as a STEM educator.

Along the way, Mica faced challenges that many young women in STEM will recognize, such as self-doubt, underrepresentation, and the pressure to prove herself. But she also discovered the power of perseverance, the importance of mentorship, and the joy of building something meaningful—whether it is a machine, a classroom, or a student's confidence.

This guide shares Mica's most valuable insights for aspiring students, especially girls and young women who are curious about STEM but unsure where to start. Her advice is rooted in real experience, shaped by resilience, and driven by a belief that everyone has a place in STEM—and that the world needs more voices like yours. In summary:

1. **Embrace the Unexpected**

Mica's journey began with a plan to become an English teacher, not an engineer. Her advice is to stay open to new experiences and let curiosity guide you. Sometimes, the most fulfilling paths are the ones you did not originally plan for.

Advice: *"Don't be afraid to pivot. Your interests can evolve, and that's not a failure—it's growth."*

2. **Persevere Through Doubt**

Whether it was the temptation to leave her program or the challenges of being a woman in a male-dominated field, Mica faced moments of uncertainty. She encourages students to push through self-doubt and seek support when needed.

Advice: *"You belong in STEM—even if you're the only one who looks like you in the room. Keep going."*

3. **Seek Mentorship and Be One**

Mentors played a crucial role in Mica's development, from her FIRST Robotics coach to her college advisor. She now pays it forward by mentoring students and coaching robotics teams.

Advice: *"Find people who believe in you—and be that person for someone else."*

4. **Use Your Engineering Mindset Everywhere**

Mica's engineering technology background taught her how to solve problems, manage systems, and think critically—skills she now uses in education, curriculum design, and leadership.

Advice: *"Engineering isn't just about machines. It's about solving problems and improving lives. That mindset is powerful anywhere."*

5. **Advocate for Yourself and Others**

Having navigated toxic work environments and the pressure to constantly prove herself, Mica emphasizes the importance of self-advocacy and creating inclusive spaces for others.

Advice: *"Speak up. Advocate for yourself and for those who haven't found their voice yet."*

6. **Stay Curious and Keep Learning**

Even after transitioning into education, Mica continues to pursue new qualifications and is considering a PhD. Her journey underscores the value of lifelong learning.

Advice: *"Never stop learning. Your education doesn't start or end with a degree—it evolves with you."*

7. **Make STEM Human**

Mica blends her love for literature and STEM to create a classroom that values creativity and logic. She shows students that STEM is not just technical—it is personal, emotional, and deeply human.

Advice: *"STEM is for storytellers, artists, and dreamers too. There's room for all of you here."*

Through Fire and Formulas: Building a Life in Engineering

Salena Nadjarian

Introduction

Born in California and raised in Lebanon, Selena Nadjarian has always lived between worlds, cultures, disciplines, and expectations. A Lebanese-Armenian mechanical engineering student who once dreamed of becoming a doctor now designs tools for learning. She believes that empathy is just as important in engineering as equations. This chapter shares her journey, from surviving years of bullying to mentoring future engineers, from enduring a war semester to building a dam simulation that teaches hydroelectricity. It is a story of persistence, identity, and the quiet power of believing you belong, even when the world tries to say otherwise.

S. Nadjarian (✉)
Purdue University, West Lafayette, IN, USA
e-mail: snadjari@purdue.edu

A. Lucietto (ed.), *Innovating for Tomorrow: Engineering Technology*, Women in Engineering and Science, https://doi.org/10.1007/978-3-032-07645-8_21

A Curious Mind at the Crossroads

At the end of high school, Selena found herself at a crossroads. She was passionate about literature, biology, chemistry, history, math, and physics. Following her family's encouragement, she enrolled in college on the biology track, intending to pursue medicine. But the memorization-heavy nature of biology left her uninspired. In contrast, physics and math sparked something deeper in her: a sense of logical reasoning, clarity, and discovery. Switching to mechanical engineering gave her not only academic fulfillment but a career path that aligned with her identity.

Heritage and Hardship

Selena's roots trace back to survivors of the Armenian Genocide—her ancestors who fled to Lebanon and Syria with little more than resilience. In 2003, amid renewed conflict in Lebanon, her mother fled to California, where Selena was born. That dual citizenship would later become both a tether and a tool.

Raised in Lebanon, Selena enjoyed a fairly comfortable childhood, interrupted by her parents' separation and nearly 7 years of persistent bullying in school. The social and verbal cruelty left scars, but it also forged an inner strength. Throughout it all, her younger sister, now a medical imaging student, remained a beacon of joy and support. Watching her sister follow a path Selena once considered reminded her of how far they had both come.

Engineering Education and the Search for Impact

As a mechanical engineering undergraduate, Selena pursued internships and scolarships, but her U.S. citizenship often posed barriers in Lebanon. Eventually, she secured a U.S.-based internship focused on engineering education. Initially seen as a backup, the experience reshaped her understanding of engineering's purpose.

One of her biggest insights came from noticing the gap between children's curiosity, about how bridges stand or how dams function, and their understanding of how to become the people who build them. Selena began developing interactive learning modules using QR codes and gamified simulations. In one project, students could control water flow through a dam to light a bulb, teaching them about hydroelectric power. Through this work, she realized education could be intuitive, immersive, and even beautiful.

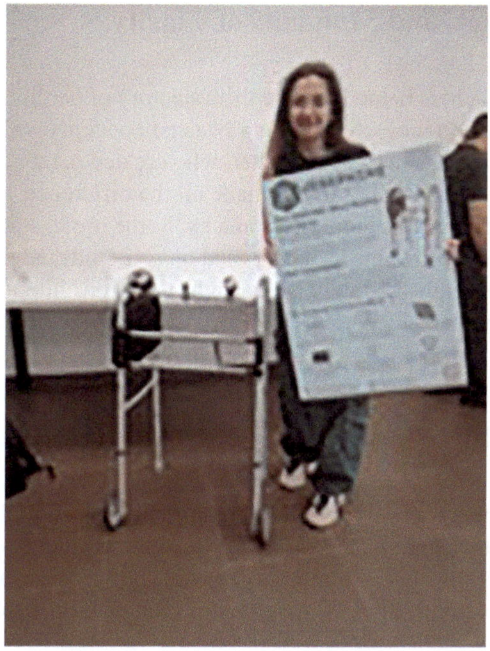

Redesigning for Dignity

One project touched her personally more than the rest. In a product development course, she designed a walker inspired by her late grandmother. Traditional models, designed for Western homes, were bulky and incompatible with Lebanon's narrower hallways. Her grandmother hated using hers.

So, Selena created a horizontally adjustable walker that respected local architecture and cultural preferences. More than a product, it was a tribute. When people asked if it would be manufactured, she hesitated. She still wonders if it should be.

Facing War and Finding Help

During one particularly difficult semester, war broke out in Lebanon. Selena attended online classes while jets screamed overhead and bombs rattled the windows. Fear was constant, and exhaustion ran deep. Eventually, she sought therapy, something rarely spoken about in her culture.

Therapy gave her a vocabulary for the stress she had silently carried for years: the bullying, the family pressure, and the academic weight. Slowly, she learned to name her emotions rather than just endure them.

Mentors, Models, and Moments of Clarity

Two high school teachers helped change the trajectory of her life. One was her physics teacher, who spent recess discussing advanced concepts with her when she had no friends. The other was her math teacher, who created personalized proofs just to match her curiosity. Their belief in her made all the difference.

In college, validation was harder to come by. Some professors dismissed her due to her small-school background. Selena relied on self-study, Khan Academy videos, trial and error, long nights. Slowly, she caught up. More importantly, she realized that her love for problem-solving was not tied to performance. It was real.

Quiet Leadership

Later in her academic journey, Selena became a mentor to younger students, especially women. Many came to her with fears: "Is engineering too hard? Is it for us?" She answered honestly: yes, engineering is hard, but that is why it needs women. Not just for gender balance but for diversity of thought: for empathy, structure, and clarity.

She noticed that all-female teams often excelled. Their logic was sound, their documentation precise, and their collaboration smooth. That was no coincidence.

When professors were dismissive or sarcastic, she encouraged her peers to speak up. "Ask him, what's funny?" she would say. Directness was her quiet form of resistance. She chose to be the kind of mentor she once needed.

Mindfulness in the Margins

What keeps Selena going is simple: she loves building things. On the first week of her internship, her professor told her to take it slow. But she could not sit still. "Please let me do something," she asked. He laughed and said, "You're the best intern ever."

That relentlessness defines her. She iterates, she stays late, she solves problems until they give. But she is also learning to pause. Walks, presence, noticing the details: these are her new tools. Through therapy, mindfulness became not a trend but a practice. She notices things others miss. She breathes deeper. It is not easy, but it is real.

Looking Forward

Graduate school is a dream she still hopes to achieve, if she can find a way to fund it. In Lebanon, there are no student loans. Students either pay or pause their education. Selena works not just for the experience but for a chance to shape the future she envisions.

One day, she hopes that students, especially young women, choose engineering not out of necessity but out of love. She hopes they enter STEM because they want to build, innovate, and lead. Because they see a place for themselves. Because someone once told them, "You belong here."

Closing

Selena does not have all the answers. But she knows how to ask the right questions. She knows how to keep showing up, even when the world feels heavy. Like engineering itself, her life is a balance of persistence, clarity, and compassion. She does not know exactly where she will end up, but she knows what she will carry with her: heart, humility, and a quiet certainty that she can figure it out.

Closing Note

Anne Lucietto

As we turn the final pages of this volume, we reflect not on an ending but on a beginning of continued dialogue, deeper understanding, and collective action. This book has brought together diverse voices and experiences, illuminating women's evolving role in engineering technology. Through research, personal narratives, and critical reflection, we've explored the persistent challenges and remarkable contributions shaping this dynamic field.

The stories shared here are more than individual journeys; they are blueprints for change. They remind us that engineering technology is not just about systems and solutions, but about people: their creativity, resilience, and courage to lead.

To every woman who has felt unseen, unheard, or underestimated this book is for you. Your presence matters. Your work matters. And your voice is shaping the future. We extend our deepest gratitude to the contributors, reviewers, and readers. May this work inspire progress, equity, and innovation, for all.

A. Lucietto
Purdue University, West Lafayette, IN, USA
e-mail: lucietto@purdue.edu

A. Lucietto (ed.), *Innovating for Tomorrow: Engineering Technology*, Women in Engineering and Science, https://doi.org/10.1007/978-3-032-07645-8

Index

A. Lucietto (ed.), *Innovating for Tomorrow: Engineering Technology*, Women in Engineering and Science, https://doi.org/10.1007/978-3-032-07645-8

MIX
Papier aus verantwortungsvollen Quellen
Paper from responsible sources
FSC® C105338

If you have any concerns about our products,
you can contact us on
ProductSafety@springernature.com

In case Publisher is established outside the EU,
the EU authorized representative is:
Springer Nature Customer Service Center GmbH
Europaplatz 3, 69115 Heidelberg, Germany

Printed by Libri Plureos GmbH
in Hamburg, Germany